Essential Mathematical Methods

Essential Mathematical Methods

Zach Short

NY RESEARCH PRESS
New York

Published by NY Research Press
118-35 Queens Blvd., Suite 400,
Forest Hills, NY 11375, USA
www.nyresearchpress.com

Essential Mathematical Methods
Zach Short

International Standard Book Number: 978-1-64725-433-9 (Hardback)

Cataloging-in-Publication Data

Essential mathematical methods / Zach Short.
p. cm.
Includes bibliographical references and index.
ISBN 978-1-64725-433-9
1. Mathematics. 2. Mathematics--Methodology. 3. Applied mathematics. I. Short, Zach.
QA36 .E87 2023
510--dc23

Contents

Preface

The world is advancing at a fast pace like never before. Therefore, the need is to keep up with the latest developments. This book was an idea that came to fruition when the specialists in the area realized the need to coordinate together and document essential themes in the subject. That's when I was requested to be the editor. Editing this book has been an honour as it brings together diverse authors researching on different streams of the field. The book collates essential materials contributed by veterans in the area which can be utilized by students and researchers alike.

Mathematical methods are widely used in various fields of science and engineering. The study of mathematical methods involves the study of multiple concepts such as advanced calculus, differential equations, complex analysis, linear algebra, and introductory mathematical physics. Linear algebra is a mathematical method that involves the study of vectors and matrices. This method has many applications ranging from mathematical physics to modern algebra and coding theory. Complex analysis or theory of functions of a complex variable is another mathematical method that studies the functions of complex numbers. Complex analysis plays a crucial role in many branches of mathematics such as algebraic geometry, number theory, analytic combinatorics, and applied mathematics. This book offers a comprehensive understanding of the essential mathematical methods. Its extensive content provides the students with a thorough understanding of various mathematical methods.

Each chapter is a sole-standing publication that reflects each author´s interpretation. Thus, the book displays a multi-facetted picture of our current understanding of application, resources and aspects of the field. I would like to thank the contributors of this book and my family for their endless support.

Zach Short

1

Interpolation and Curve Fitting

1.1 Introduction

If we consider the statement $y = f(x), x_0 \le x \le x_n$ we understand that we can find the value of y, corresponding to every value of x in the range $x_0 \le x \le X_n$.

Definition: If the function f(x) is single valued and continuous and is known as explicitly then the values of f(x) for certain values of x like x_0, x_1, x_2, ..., xn can be calculated. The problem now is if we are given the set of tabular values,

X	x_0	x_1	x_1	....	x_{n-2}	x_{n-1}	x_n
Y	y_0	y_1	y_2	....	y_{n-2}	y_{n-1}	y_n

Satisfying the relation y = f(x) and the explicit definition of f(x) is not known, is it possible to find a simple function say (x) such that f(x) and ϕ(x) agree at the set of tabulated points. This process of finding (x) is called "Interpolation".

Definition: If ϕ (x) is a polynomial then the process is called "polynomial interpolation" and ϕ (x) is called "interpolating polynomial".

Errors in Polynomial Interpolation

Types of Errors: Let x be the value and x* be an approximation to the value x. Then the difference x – x* is called an "Error" in x*.

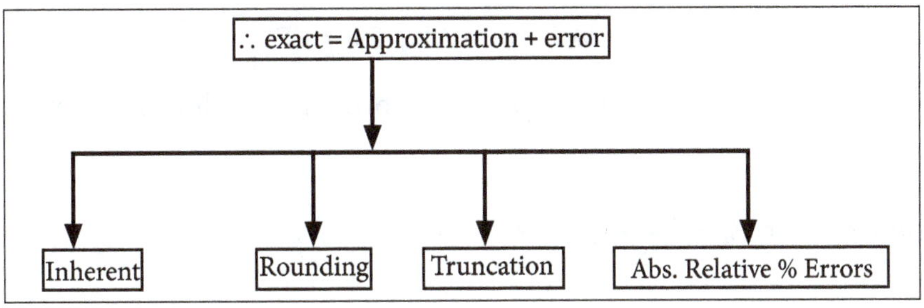

(i) Inherent Errors: Errors which are already involved in the statement of a problem before its solution are called "Inherent Errors". These types of errors arise either due to the given data being approximate or due to the limitations of math tables calculations or the digital computer.

Inherent errors can be minimized by taking better date or by using precision computing skills or aids.

(ii) Rounding errors: Errors which arise from the process of rounding off the number during the calculations are known as "rounding error".

Let two numbers each having n digits be multiplied and the resulting number which will contain 2n digits be rounded to "n" digits. In doing so, "a round – off error is introduced".

Similarly, in a decimal system, there may be infinity of digits to the right of the decimal point, and it may not be possible for us to use infinity of digits in a computational problems. So, we use only finite number of digits only in our calculations.

This is the source of the so called "Rounding of Errors". Rounding errors can be reduced by:

- Changing the calculation procedure or division of a small number.
- Retaining at least one or more significant figure at each step than that given in the data and rounding off at the last step.

(iii) Truncation Errors: These errors arise due to approximation of results or on replacing an infinite process by a finite one.

Example: if $\sin x = x - \frac{x^3}{3!} + \frac{x^5}{5!} - \frac{x^7}{7!} + ...$ in this $\frac{x^3}{3!} + \frac{x^5}{5!} - \frac{x^7}{7!} + ...$ is replaced by x' then $\sin x = x - x'$.

Truncation error type is algorithm error.

The magnitude of the error in the value of the function due to cutting (truncation) of its series is equal to the sum of all the discarded terms.

It may be large and may even exceed the sum of the terms retained, thus making calculated result meaningful.

(iv) Absolute Error or Relative or % Errors: AE / RE / PE

If x is the value of a quantity and x* is the its approximate value and then $|x - x^*|$ is called "Absolute error E_o".

$$\text{Relative error: } E_r = \left|\frac{x - x^*}{x}\right|$$

$$\text{\% Error: } E_p = 100E_r = \left|\frac{x - x^*}{x}\right|$$

If x such that $|x - x^*| = x$ then x is an upper limit on the magnitude of absolute error and measures the absolute accuracy.

Note: The relative error, % errors are independent and absolute error is the expression in terms of E_r, E_p of these units.

If a number is correct to n decimal places, then the error is $\frac{1}{2}.10^{-n}$

If the first significant figure of a number is k and the number is correct to n significant figures, then the relative error $E_r < \frac{1}{k.x.10^{n-1}}$

Definition: Errors in the Approximation of a Function

Let $y = f(x_1, x_2)$ for all x_1, x_2.

If $\delta x_1, \delta x_2$ be the errors in x_1, x_2 then the error δy in y is given by:

$y + \delta y = \delta(x_1 + \delta x_1, x_2 + \delta x_2) = f(x_1, x_2) + \left(\frac{\partial f}{\partial x_1}.\delta x_1 + \frac{\partial f}{\partial x_2}.\delta x_2\right) +$ terms involving high power of $\delta x_1 \delta x_2$ are infinitesimal their squares and higher power can be neglected.

Then, $y + \delta y = f(x_1 + x_2) + \left(\frac{\partial f}{\partial x_1}.\delta x_1\right) + \left(\frac{\partial f}{\partial x_2}.\delta x_2\right) \Rightarrow \delta y = \left(\frac{\partial f}{\partial x_1}.\delta x_1\right) + \left(\frac{\partial f}{\partial x_2}.\delta x_2\right)$

In general, δy in $y = f(x_1, x_2, x_3, x_4, \ldots . x_n)$ corresponding to error δx_1 in X_i for i = 1,2,3,….,n given by $\left(\delta y = \frac{\partial y}{\partial x_1}.\delta x_1\right) + \left(\frac{\partial y}{\partial x_2}.\delta x_2\right) + \ldots . . + \left(\frac{\partial y}{\partial x_n}\delta x_n\right)$

Relative error is $E_r = \frac{\partial y}{y}\left(\frac{\partial y}{\partial x_1}.\frac{\delta x}{y}\right) + \left(\frac{\partial y}{\partial x_2}.\frac{\delta x_2}{y}\right) + \ldots . . + \left(\frac{\partial y}{\partial x_n}.\frac{\delta x_n}{y}\right)$

1.2 Finite Differences

We noted that N^{th} degree accurate Finite Difference (FD) expressions for first derivatives have an associated error.

$$E \cong h^N \frac{d^{N+1}f}{dx^{N+1}}$$

- If f(x) is an Nth degree polynomial then the FD approximation to the first derivative is exact.

$$\frac{d^{N+1}f}{dx^{N+1}} = 0$$

- Thus if we know that a FD approximation to a polynomial function is exact, we can derive the form of that polynomial by integrating the previous equation.

This implies that a distinct relationship exists between polynomials and FD expressions for derivatives (different relationships for higher order derivatives).

$$f(x) \cong a_1 x^N + a_2 x^{N-1} + \ldots a_{N+1}$$

- We can in fact develop FD approximations from interpolating polynomials, Approximations to First and Second Derivatives Using Quadratic Interpolation.
- We will illustrate the use of interpolation to derive FD approximations to first and second derivatives using a 3 node quadratic interpolation function.

For first derivatives p=1 we must establish at least an interpolating polynomial of degree N=1 with N+1=2 nodes.

For second derivatives p=2 we must establish at least an interpolating polynomial of degree N=2 with N+1=3 nodes.

Thus a quadratic interpolating function will allow us to establish both first and second derivative approximations.

- Apply a shifted coordinate system to simplify the derivation without affecting the generality of the derivation.

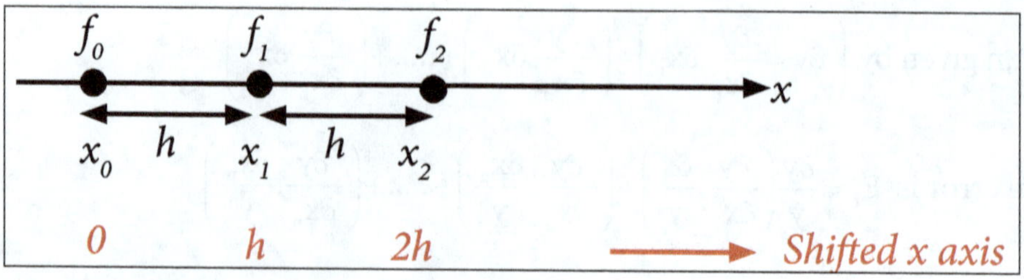

Problems

1. Let us convert the differential equation $y''(x) + y'(x) + y = 0$ into finite difference equivalent form.

Solution:

The finite difference equation is:

$$\frac{y_{i-1} - 2y_i + y_{i+1}}{h^2} + \frac{y_{i+1} - y_i - 1}{2h} + y_i = 0$$

Multiply both sides by $2h^2$:

$$2y_{i-1} - 4y_i + 2y_{i+1} + hy_{i+1} - hy_{i-1} + 2h^2 y_i = 0$$

$$(2-h)y_{i-1}+(2h^2-4)y_i+(h+2)y_{i+1}=0$$

2. Let us write down the finite difference for the following second order ODE with $h=1/n, Y''=y+x, y(0)=y(1)=0$.

Solution:

Given: $y''(x)-y(x)=x$

$$\frac{y_{i-1}-2y_i+y_{i+1}}{h^2}-y_i=x_i$$

i.e., $n^2\, y_{i-1}(2n^2+1)y_i.+n^2\, y_{i+1}=x_i$

3. Let us solve the BVP $\frac{d^2y}{dxz}-y=0$ with $y(0)=0, y(1)=1$, using finite difference method with h = 0.2.

Solution:

Given: $y''-y=0$ and $h=0.2$

$$x_0=0;\; y_0=0$$

x	$x_0 = 0$	$x_1 = 0.2$	$x_2 = 0.4$	$x_3 = 0.6$	$x_4 = 0.8$	$x_5 = 1$
y	$y_0 = 0$	y_1	y_2	y_3	y_4	$y_5 = 1$

The difference equation for $y''-y=0$ is:

$$\frac{y_{i+1}-2y_i+y_{i-1}}{h^2}-y_i=0. \text{ i.e. } y_{i+1}(-2+h^2)y_i+y_{i-1}=0$$

$$h=0.2;\; i=1,2,3,4$$

$$y_{i+1}-2.04\,y_i+y_{i-1}=0$$

Put $i=1,2,3,4$

$$y_2-2.04\,y_1+y_0=0$$

i.e. $y_2-2.04\,y_1=0$ ...(1)

$y_3-2.04y_2+y_1=0$ ...(2)

$y_4-2.04y_3+y_2=0$ ...(3)

$$y_5 - 2.04y_4 + y_3 = 0$$

$$1 - 2.04\,y_4 + y_3 = 0$$

i.e. $-2.04y_4 + y_3 = -1$ ...(4)

Solving (1), (2), (3) and (4):

$(2) \times 2.04 \Rightarrow 2.04y_3 - 4.16l6y_2 + 2.04y_1 = 0$ ...(5)

$(1)+(2) \Rightarrow 2.04y_3 - 3.1616y_2 = 0$ ...(6)

$(3) \times 2.04 \Rightarrow 2.04y_4 - 4.1616y_3 + 2.04y_2 = 0$ ...(7)

$(7)+(4) \Rightarrow -3.1616y_3 + 2{,}04y_2 = -1$ ...(8)

$(6) \times 2.04 \Rightarrow 4.1616y_3 - 6.4497y_2 = 0$ ...(9)

$(8) \times 3.1616 \Rightarrow -9.9957y_3 + 6.4497y_2 = -3.1616$ ...(10)

$(9)+(10) \Rightarrow -5.8341y_3 = -3.1616$

i.e. $y_3 = 0.5419$

Substitute in (4), we get $y_4 = 0.7558$

Substitute in (3), we get $y_2 = 0.3497$

From (1), we get $y_1 = 0.1714$

$$y(0.2) = 0.1714 . y(0.4) = 0.3497$$

$$y(0.6) = 0.5419 y(0.8) = 0.7558.$$

4. Let us solve the BVP y″+ y = 0, y(0) = 1, y(1) = 0 using finite difference method, taking h = 0.25.

Solution:

The given differential equation can be written as:

$y''(x) + y(x) = 0$ ...(1)

i.e., $\frac{y_{i-1} - 2y_i + y_{i+1}}{h^2} + y_i = 0$

$y_{i-1} + \left(h^2 - 2\right)y_i + y_i + 1 = 0$ ...(2)

x	0	0.25	0.5	0.75	1.
y	1				0
	y_0	y_1	y_2	y_3	Y_4

To find: $y_1 = y(0.25)$, $y_2 = y(0.5)$ & $y_3 = y(0.75)$

Here h = 0.25

Put i = 1 in (2):

$$y_0 + (0.0625 - 2)y_1 + y_2 = 0$$

$$1 - 1.9375\, y_1 + y_2 = 0$$

i.e., $1.9375\, y_1 - y_2 = 1$ ...(3)

Put i = 2 in (2):

$$y_1 - 1.9375\, y_2 + y_3 = 0 \quad ...(4)$$

Put i = 3 in (2)

$$y_2 - 1.9375\, y_3 + y_4 = 0$$

i.e., $y_2 - 1.9375\, y_3 = 0$ ...(5)

Solving (3), (4) and (5) we get:

$$y_1 = y(0.25) = 0.8104$$

$$y_2 = y(0.5) = 0.5702$$

$$y_3 = y(0.75) = 0.2943$$

5. Using finite different let us solve the boundary value problem y” + 3y’ – 2y = 2x + 3,.y (0) = 2, y (1) = 1 with h = 0.2.

Solution:

The given different equation can be written as:

$$y''(x) + 3y'(x) - 2y\ (x) = 2.x + 3 \quad ...(1)$$

Using the central different approximation

$$\therefore\ y^n = \frac{y_{k-1} - 2y_k + y_k + 1}{h^2} \quad ...(2)$$

$$y^{n} = \frac{y_{k+1} - y_{k} - 1}{2h} \qquad ...(3)$$

Sub (2) & (3) in (1):

$$\therefore \frac{y_{k-1} - 2y_{k} + y_{k} + 1}{h^{2}} + \frac{3}{2}\left[\frac{y_{k+1} - y_{k} + y_{k} - 1}{h}\right] - 2y_{k} = 2x_{k} + 3$$

$$2\left[y_{k-1} - 2y_{k} + y_{k+1}\right] + 3h\left[y_{k+1} - y_{k-1}\right] - 4h^{2}y_{k} = \left(2x_{k} + 3\right)2h^{2}$$

$$2y_{k-1} - 4y_{k} + 2y_{k} + 1 + \frac{3}{5}y_{k-1} - \frac{4}{25}y_{k} = \frac{2}{25}\left(2x_{k} + 3\right)$$

$$50y_{k-1} - 100y_{k} + 50y_{k+1} + 15y_{k+1} - 15y_{k-1} - 4y_{k} = 4x_{k} + 6$$

$$35y_{k-1} - 104y_{k} + 65y_{k+1} = 4x_{k} + 6$$

Put $k = 1,2,3,4$

$$\therefore \left.\begin{aligned} 35y_{0} - 104y_{1} + 65y_{2} &= 4x_{1} + 6 \\ 35y_{1} - 104y_{2} + 65y_{3} &= 4x_{2} + 6 \\ 35y_{2} - 104y_{3} + 65y_{4} &= 4x_{3} + 6 \\ 35y_{3} - 104y_{4} + 65y_{5} &= 4x_{4} + 6 \end{aligned}\right\} \qquad ...(A)$$

$$\left.\begin{aligned} \text{Given}: y(0) = 2 \text{ ie., } & x_{0} = 0,\ y_{0} = 2 \\ y(1) = 1 \qquad & x_{5} = 1,\ y_{5} = 1 \end{aligned}\right\} \qquad ...(B)$$

Since $h = \frac{1}{5}$, we have:

$$x_{1} = \frac{1}{5}, x_{2} = \frac{2}{5}, x_{3} = \frac{3}{5}, x_{4} = \frac{4}{5}$$

Sub (B) in (A)

$$\therefore \quad 35(2) - 140y_{1} + 65y_{2} = \frac{4}{5}$$

$$-104y_{1} + 65y_{2} = \frac{34}{5} - 70$$

$$-104y_1 + 65y_2 = -63.2 \quad ...(4)$$

$$35y_1 - 104y_2 + 65y_3 = 2\left(\frac{2}{5}\right) + 6$$

$$35y_1 - 104y_2 + 65y_3 = 7.6 \quad ...(5)$$

$$35y_2 - 104y_3 + 65y_4 = 4\left(\frac{3}{5}\right) + 6$$

$$35y_2 - 104y_3 + 65y_{34} = 8.4$$

$$35y_3 - 104y_4 + 65(1) = 4\left(\frac{4}{5}\right) + 6$$

$$35y_2 - 104y_3 + 65y_{34} = 8.4 \quad ...(6)$$

$$35y_3 - 104y_4 = -55.8 \quad ...(7)$$

$$-104y_1 + 65y_2 = -63.2 \quad ...(4)$$

$$35y_1 - 104y_2 + 65y_3 = 7.6 \quad ...(5)$$

$$(4) \times 35 \Rightarrow -3640y_1 + 2275v_2 = -2212 \quad ...(8)$$

$$(5) \times 104 \Rightarrow 3640y_1 - 10816y_2 + 6760y_3 = 790.4 \quad ...(9)$$

$$(8) + (9) \Rightarrow -8541y_2 + 6760y_3 = -1421.6 \quad ...(10)$$

$$35y_2 - 104y_3 + 65y_4 = 8.4 \quad ...(6)$$

$$35y_3 - 104y_4 = -55.8 \quad ...(7)$$

$$(6) \times 104 \Rightarrow 3640y_2 - 10816y_3 + 6760y_4 = 873.6 \quad ...(11)$$

$$(7) \times 65 \Rightarrow 2275y_3 - 6760y_4 = -3627 \quad ...(12)$$

$$(11) + (12) \Rightarrow 3640y_2 - 8541y_3 = -2753.4 \quad ...(13)$$

$$(10) \times 8541 \Rightarrow -72948681y_2 + 57737160y_3 = -12141885.6 \quad ...(14)$$

$$(13) \times 6760 \Rightarrow 24606400y_2 - 57737160y_3 = -18612984$$

$$-48342281y_2 = -30754869.6 \quad ...(15)$$

$y_2 = 0.6362$

Substitute: $y_2 = 0.6362$ in (4)

$\Rightarrow -104y_1 + 65\,(0.6362) = -63.2$

$-104y_1 = -104.553$

$y_1 = 1.005$

Substitute: $y_1 = 1.005,\ y > y_2 - 0.6362$ in (5)

$\Rightarrow\ 35(1.005) - 104(0.6362) + 65y_3 = 7.6$

$y_3 = 0.5937$

Substitute: $y_3 = 0.5937$ in (7)

$\Rightarrow 35(0.5937) - 104y_4 = -55.8$

$y_4 = 0.7363$

$\therefore\ y_1 = 1.005;\ y_2 = 0.6362$

$y_3 = 0.5937;\ y_4 = 0.7363$

1.2.1 Forward Differences

Newton's forward interpolation formula is used to interpolate the values of y near the beginning of the tabular values.

If $y_o, y_1, y_2 \ldots\ldots\ldots\ldots y_n$ denote a set of values of y, then the first forward differences of y=f(x) are defined by $\Delta y_o = y_1 - y_o; \Delta y_1 = y_2 - y_1; \ldots. \Delta y_{n-1} = y_n - y_{n-1}$.

Where Δ is called the forward difference operator.

Problems

1. Let us find the value of y when x = 5 using Newton's interpolation formula from the following table:

x:	4	6	8	10
y:	1	3	8	16

Solution:

Difference Table:

X	Y	Δy		$\Delta^2 y$		$\Delta^3 y$	
$4x_0$	$1y_0$		$\Delta^2 y_0$				
		3 - 1 = 2			$\Delta^2 y_0$		
				5 - 2 = 3			$\Delta^3 y_0$
6	3					3 - 3 = 0	
		8 - 3 = 5					
				8 - 5 = 3			
8	8						
		16 - 8 = 8					
10	16						

Newton's forward interpolation formula is:

$$y(x) = y_0 + \frac{u}{1!}\Delta y_0 + \frac{u(u-1)}{2!}\Delta^2 y_0 + \frac{u(u-1)(u-2)}{3!}\Delta^3 y_0 + \ldots$$

Where, $u = \dfrac{x - x_0}{h}$

Here $h = 2$.

$$u = \frac{x-4}{2}$$

$$y(x) = 1 + \left(\frac{x-4}{2}\right)(2) + \frac{1}{2}\left(\frac{x-4}{2}\right)\left(\frac{x-6}{2}\right)(3) + 0\ldots$$

$$= 1 + x - 4 + \frac{3}{8}\left(x^2 - 10x + 24\right).$$

$$y(x) = \frac{3}{8}x^2 - \frac{11}{4}x + 6$$

$$y(5) = \frac{3}{8}(25) - \frac{11}{4}(5) + 6 = 1.625.$$

2. Using Newton's Forward interpolation formula, let us find f (1.02) from the following data.

X	1.0	1.1	1.9	1.3	1.4
F(x)	0.841	0.891	0.932	0.964	0.985

Solution:

Forward difference table:

x	y = f(x)	Δy	$\Delta^2 y$	$\Delta^3 y$
1.0	0.841			
1.1	0.891	0.050		
1.2	0.932	0.041	−0.009	
1.3	0.964	0.032	−0.009	0
1.4	0.985	0.021	−0.011	−0.002

$$y_0 = 0.841, \Delta y_0 = 0.050, \Delta^2 y_0 = -0.009$$

Let $x_0 = 1.0 \,\&\, x = 1.02$

$$p = \frac{x - x_0}{h}$$

$$\Rightarrow p = 0.2$$

By Newton's Forward interpolation formula:

$$y(1.02) = y_0 + p\Delta y_0 + \frac{p(p-1)}{2!}\Delta^2 y_0 + \ldots$$

$$= 0.852$$

$$y(1.02) = 0.852$$

1.2.2 Backward Differences

Newton's backward interpolation formula is used to interpolate the values of y near the end of a set of tabular values.

If $y_1 - y_0, y_2 - y_1 \ldots\ldots y_n - y_{n-1}$ denote a set of values of y, then the first forward differences of y=f(x) are defined by $\nabla y_1 = y_1 - y_0; \nabla y_2 = y_2 - y_1; \ldots\ldots \nabla y_n = y_n - y_{n-1}$.

Where ∇ is called the backward difference operator.

Problems

1. Using Newton's Backward interpolation formula, let us find y when x=27, from the following data.

x	10	15	20	25	30
y	35.4	32.2	29.1	26.0	23.1

Solution:

Backward difference table:

x	y	∇y	$\nabla^2 y$	$\nabla^3 y$	$\nabla^4 y$
10	35.4				
		−3.2			
15	32.2		0.1		
		−3.1		−0.1	
20	29.1		0		0.3
		−3.1		0.2	
25	26.0		0.2		
		−2.9			
30	23.1				

Here $x_n = 30, y_n = 23.1, \nabla y_n = 2.9, \nabla^2 y_n = 0.2, \nabla^4 y_n = 0.3$

By Newton's backward interpolation formula is:

$$y(x_n + ph) = y_n + p\nabla y_n + \frac{p(p+1)}{2!}\nabla^2 y_n + \frac{p(p+1)(p+2)}{3!}\nabla^3 y_n + \ldots$$

$$\text{Here } x = 27, p = \frac{x - x_n}{h}$$

$$\Rightarrow p = -0.6$$

$$y(27) = 24.8$$

1.2.3 Central Differences

Let us consider $x_0, x_1, x_2, \ldots x_n$ be given set of observation and let $y_0, y_1, y_2 \ldots y_n$ are corresponding values of the curve $y = f(x)$ then the central difference operator is denoted by δ and is defined as,

If n is odd: $\delta^n y_{r-\frac{1}{2}} = \delta^{n-1} y_{r-1,}\ r = 1,2,3,\ldots$

If n is even: $\delta^n y_r = \delta^{n-1} y_{r+\frac{1}{2}} - \delta^{n-1} y_{r-\frac{1}{2}}, r = 1,2,3,\ldots$

and $\delta y_r = y_r$

x	y	δy	$\delta^2 y$	$\delta^3 y$	$\delta^4 y$
x_0	y_0	$\delta y_{\frac{1}{2}}$			
x_1	y_1		$\delta^2 y_1$		
		$\delta y_{\frac{3}{2}}$		$\delta^3 y_{\frac{3}{2}}$	
x_2	y_2		$\delta^2 y_2$		$\delta^4 y_2$
		$\delta y_{\frac{5}{2}}$			
x_3	y_3				
			$\delta^2 y_3$		
x_4	y_4	$\delta y_{\frac{7}{2}}$			

Let h be common difference in the values of x and y = f (x) be given function then:

$$\delta f(x) = f\left(x+\frac{h}{2}\right) - f\left(x-\frac{h}{2}\right)$$

1.2.4 Symbolic Relations and Separation of Symbols

Difference formulae can easily be established by symbolic methods, using the shift operator E and the averaging or the mean operator μ, in addition to the operators Δ and δ already defined.

The averaging operator μ is defined by the equation:

$$\mu y_r = \frac{1}{2}\left(y_{r+1/2} + y_{r-1/2}\right)$$

The shift operator E is defined by the equation:

$$Ey_r = y_{r+1},$$

Which shows that the effect of E is to shift the functional value yr to the next higher value yr+1. A second equation with E gives:

$$E^2 y_r = E(Ey_r) = Ey_{r+1} = y_{r+2}$$

And in general:

$$E^n y_r = y_{r+n}$$

It is now easy to derive a relationship between Δ and E, for we have:

$$\Delta y_o = y_1 - y_o = Ey_o - y_o = (E-1)y_o$$

and hence:

$$\Delta \equiv E-1 \quad \text{or} \quad E \equiv 1+\Delta$$

We can now express any higher-order forward difference in terms of the given function values. For example:

$$\Delta^3 y_o = (E-1)^3 y_o = (E^3 - 3E^2 + 3E - 1)y_o = y_3 - 3y_2 + 3y_1 - y_o$$

From the definitions, the following relations can easily be established:

$$\nabla = 1 - E^{-1}$$

$$\delta = E^{1/2} - E^{-1/2}$$

$$\mu = (1/2)(E^{1/2} + E^{-1/2}), \mu^2 = 1 + (1/4)\delta^2$$

$$\Delta = \nabla E = \delta E^{1/2}$$

As an example, we prove the relation:

$$\mu^2 \equiv 1 + (1/4)\delta^2$$

We have, by definition:

$$\mu y_r = \frac{1}{2}\left(y_{r+1/2} + y_{r-1/2}\right)$$

$$= \frac{1}{2}\left(E^{1/2} y_r + E^{-1/2} y_r\right)$$

$$= \frac{1}{2}\left(E^{1/2} + E^{-1/2} y_r\right)$$

Hence:

$$\mu = \frac{1}{2}\left(E^{1/2} + E^{-1/2}\right)$$

And:

$$\mu^2 = \frac{1}{4}\left(E^{1/2} + E^{-1/2}\right)^2$$

$$=\frac{1}{4}\left(E+E^{-1}+2\right)$$

$$=\frac{1}{4}\left[\left(E^{1/2}+E^{-1/2}+2\right)^2+4\right]$$

$$=\frac{1}{4}\left(\delta^2+4\right)$$

We therefore have:

$$\mu\equiv\sqrt{1+\frac{1}{4}\delta^2}$$

Finally, we define the operator D such that:

$$Dy(x)=\frac{d}{dx}y(x)$$

To relate D to E, we start with the Taylor's series:

$$y(x+h)=Y(x)+hy'(x)+\frac{h^2}{2!}y''(x)+\frac{h^3}{3!}y'''(x)+\ldots$$

This can be written in the symbolic from:

$$Ey(x)=\left(1+hd+\frac{h^2D^2}{2!}+\frac{h^3D^3}{3!}+\ldots\right)y(x)$$

Since the series in the brackets is the expansion of e^{hd}, we obtain the interesting result:

$$E\equiv e^{hD}$$

Using the relation, a number of useful identities can be derived. This relation is used to separate the effect of E into that of the powers of A and this method of separation is called the method of separation of symbols. The following examples demonstrate the use of this method.

Example:

Using the method of separation of symbols, let's show that:

$$\Delta^n u_{x-n}=u_x-nu_{x-1}+\frac{n(n-1)}{2}u_{x-2}+\cdots+(-1)^n u_{x-n}$$

To prove this result, we start with the right-hand side. Thus:

$$u_x - nu_{x-1} + \frac{n(n-1)}{2}u_{x-2} + \cdots + (-1)^n u_{x-n}$$

$$= u_x - nE^{-1}u_x + \frac{n(n-1)}{2}E^2 u_x + (-1)^n E^{-n} u_x$$

$$= \left[1 - nE^{-1} + \frac{n(n-1)}{2}E^{-2} + \cdots + (-1)^n E^{-n}\right] u_x$$

$$= (1 - E^{-1})^n u_x$$

$$= \left(1 - \frac{1}{E}\right)^n u_x$$

$$= \left(\frac{E-1}{E}\right)^n u_x$$

$$= \frac{\Delta^n}{E^n} u_x$$

$$= \Delta^n E^{-n} u_x$$

$$= \Delta^n u_{x-n}$$

1.3 Difference Equations

Let $z(\cdot)$ and $w(\cdot)$ be sequences of real number. Then:

$$z(t) + \alpha_1 z(t-1) + \cdots + \alpha_p z(t-p) = \omega t$$

Is called a difference equation of order P, coeffients $\alpha_1, \cdots, \alpha_p$ and forcing term $w(\cdot)$. If $w(t) = o$ for all t, we say the different equation is homogeneous.

1.3.1 Differences of a Polynomial and Newton's Formulae for Interpolation

Differences of a Polynomial

Let y(x) be the polynomial of the nth degree so that:

$$y(x) = a_o x^n + a_1 x^{n-1} + a_2 x^{n-2} + \cdots + a_n$$

Then we obtain:

$$y(x+h)-y(x)=a_0\left[(x+h)^n-x^n\right]+a_1\left[(x+h)^{n-1}-x^{n-1}\right]+\cdots$$

$$=a_0\,(nh)\,x^{n-1}+a_1'x^{n-2}+\ldots+a_n'$$

Where $a_1', a_2', \cdots a_n'$ are the new coefficients.

The above equation can be written as:

$$\Delta y(x)=a_0(nh)x^{n-1}+a_1'x^{n-2}+\cdots+a_n'$$

Which shows that the first difference of a polynomial of the nth degree is a polynomial of degree (n — 1).

Similarly, the second difference will be a polynomial of degree (n — 2), and the coefficient of x^{n-2} will be $a_0n(n-1)h^2$. Thus the nth difference is $a_0n!h$", which is a constant.

Hence, the (n + 1)th, and higher differences of a polynomial of nth degree will be zero.

Conversely, if the nth differences of a tabulated function are constant and the (n + 1)th, (n + 2)th .....differences all vanish and then the tabulated function represents a polynomial of degree n. It should be noted that these results hold good only if the values of x are equally spaced. The converse is important in numerical analysis since it enables us to approximate a function by a polynomial if its differences of some order become nearly constant.

Newton's Formulae for Interpolation

If the values of x are given at unequal intervals, it is convenient to introduce the idea of divided differences. The divided differences are the differences of y=f(x) defined, taking into consideration the changes in the values of the argument.

Using divided differences of the function y=f(x), we establish Newton's divided difference interpolation formula which is used for interpolation and the values of x are at unequal intervals and also for fitting an approximate curve for the given data.

Divided Difference

Let the function y = f(x) assume the values $f(x_0)$, $f(x_1)$, $f(x_2)$...$f(x_n)$.

Corresponding to the arguments $x_1, x_2, x_3 \ldots x_n$ respectively, where the intervals $x_1-x_0, x_2-x_1 \ldots\ldots x_n-x_{n-1}$ need to be equal.

Definitions

The first divided difference of f(x) for the arguments x_0, x_1 is defined by:

$$f(x_0,x_1)=\Delta_1 f(x_0)=\frac{f(x_1)-f(x_0)}{x_1-x_0}$$

It is also denoted by $[x_0, x_1]$.

The second divided divided differences of f(x) for three arguments x_0, x_1, x_2 is defined as:

$$f(x_0,x_1,x_2)=\Delta_1{}^2 f(x_0)=\frac{f(x_1,x_2)-f(x_0,x_1)}{x_2-x_0}$$

The third divided difference of f(x) for the four arguments x_0, x_1, x_2, x_3 is defined as:

$$f(x_1,x_2,x_3)=\frac{f(x_1,x_2,x_3)-f(x_0,x_1,x_2)}{x_3-x_0}$$

And so on.

Representation by Divided Difference Table

Arguments	Entry f(x)	First divided difference $\Delta_1 f(x)$	Second D – D $\Delta_1^2 f(x)$	Third D – D $\Delta_1^3 f(x)$
x_0	$F(x_0)$			
		$f(x_0,x_1)=\Delta_1 f(x_0)=\frac{f(x_1)-f(x_0)}{x_1-x_0}$		
x_1	$F(x_1)$		$f(x_0,x_1,x_2)$	
				$f(x_0,x_1,x_2,x_3)$
x_2	$F(x_0)$	$f(x_1,x_2)=\Delta_1 f(x_1)=\frac{f(x_2)-f(x_1)}{x_2-x_1}$	$f(x_1,x_1,x_2)$	
				$f(x_1,x_2,x_3,x_4)$
x_3	$F(x_0)$	$f(x_1,x_2)=\Delta_1 f(x_1)=\frac{f(x_2)-f(x_1)}{x_2-x_1}$	$f(x_2,x_3,x_4)$	
x_4	$F(x_0)$			
		$f(x_1,x_2)=\Delta_1 f(x_1)=\frac{f(x_2)-f(x_1)}{x_2-x_1}$		

Properties of Divided Difference

- The divided differences are symmetrical in all their arguments.
- The operator Δ_1 is linear.
- The nth divided differences of a polynomial of nth degree are constants.

Problems

1. Let us find the divided differences of $f(x) = x^3 - x^2 + 3x + 8$ for the arguments 0, 1, 4, 5.

Solution:

x	y = f(x)	Δy	$\Delta^2 y$	$\Delta^3 y$
0	8	3	4	1
1	11	19	9	
4	68	55		
5	123			

2. Let us find the first and second divided differences with arguments a, b, c of the function $f(x)=\frac{1}{x}$

Solution:

x	f(x)	Δf(x)	$\Delta^2 f(x)$
a	$\left\lvert\frac{1}{a}\right\rvert$	$\frac{\frac{1}{b}-\frac{1}{a}}{b-a}=-\frac{1}{ab}$	
b	$\left\lvert\frac{1}{b}\right\rvert$	$\frac{\frac{1}{c}-\frac{1}{b}}{c-b}=-\frac{1}{bc}$	$\frac{-\frac{1}{bc}-\frac{1}{ab}}{c-a}=\frac{1}{abc}$
c	$\left\lvert\frac{1}{c}\right\rvert.$		

3. Let us find the first derivative of f(x) at x = 2 for the data f(- 1) = – 21, f(1) = 15, f(2) = 12 and f(3) = 3, using Newton's divided difference formula.

Solution:

Given:

x	−1	1	2	3
y = f(x)	−21	15	12	3

Divided Difference Table:

	x	y	Δy	$\Delta^2 y$	$\Delta^3 y$
(y_0)					
(x_0)	−1	$-21(y_0)$	(Δy_0)		
			$\frac{15-(-21)}{1-(-1)}=18$	$(\Delta^2 y_0)$	
(x_1)	1	15		−7	$(\Delta^3 y_0)$
			−3		1
(x_2)	2	12		−3	
			−9		
(x_3)	3	3			

$$y = y_0 + (x - x_0)\Delta y_0 + (x - x_0)(x - x_1)\Delta^2 y_0$$

$$+(x - x_0)(x - x_1)(x - x_2)\Delta^2 y_0 +$$

$$y = -21 + (x+1)(18) + (x + 1)(x - 1)(-7)$$

$$+ (x + 1)(x - 1)(x - 2)(1) +$$

$$= -21 + 18x + 18 - 7x^2 + 7 + x^3 - 2x^2 - x + 2$$

$$y(x) = x^2 - 9x^2 + 17x + 6$$

$$y(x) = 3x^2 - 18x + 17$$

$$y(2) = 3(4) - 18(2) + 17 = -7$$

4. Using Newton's divided difference formula, let us find f(x) from the following data and hence find f(4).

x:	0	1	2	5
f(x):	2	3	12	147

Solution:

Divided Difference Table:

X	y=f(x)	Δy	$\Delta^2 y$	$\Delta^3 y$
X_0 o	y_0			
	2	$\Delta^2 y_0$		

		$\frac{3-2}{1-0}=1$	$\Delta^2 y_0$	
x_1 1	3		$\frac{9-1}{2-0}=4$	$\Delta^3 y_0$
				$\frac{9-4}{5-0}=1$
		$\frac{12-3}{2-1}=9$		
x_2 2	12		$\frac{45-9}{5-1}=9$	
		$\frac{147-12}{5-2}=45$		
x_3 5	147			

Newton's divided difference formula is:

$$f(x)=y_0+(x-x_0\ \Delta y_0+(x-x_0)(x-x_1)\Delta^2 y_0$$

$$+(x-x_0)(x-x_1)(x-x_2)\Delta^3 y_0+\ldots$$

$$f(x)=2+(x-0)(1)+(x-0)(x-1)(4)$$

$$+(x-0)(x-1)(x-2)(1)+\ldots$$

$$=2+x+4x^2-4x+x^3-3x^2+2x$$

$$f(x)=-x^3+x^2-x+2$$

Put $x=4$

$$f(4)=4^3+4^2-4+2=78$$

5. By using Newton's divided difference formula, let us find f(8).

x	4	5	7	10	11	13
f(x)	48	100	294	900	1210	2028

Solution:

Divided Difference Table:

x	y = f(x)	Δy	Δ²y	Δ³y	Δ⁴y
x_0 4	48 y_0	$\Delta^y{}_0$		-	
		$\frac{100-48}{5-4}=52$	$\Delta^2 y_0$		
x_1 5	100		$\frac{97-52}{7-4}=15$	$\Delta^3 y_0$	
		$\frac{294-100}{7-5}=97$		$\frac{21-15}{10-4}=1$	
x_2 7	294		21		0
		202		1	
x_3 10	900		27		0
		310		1	
x_4 11	1210		33		
		409			
x_5 13	2028				

Newton's divided difference formula is:

$$y(x)=y_0+(x-x_0)\,\Delta y_0+(x-x_0(x-x_1)\Delta_2 y_0$$

$$+(x-x_0)(x-x_1)\,(x-x_2)\,\Delta^3 y_0+\ldots$$

$$f(x)=y=48+(x-4)(52)+(x-4)(x-5)(15)$$

$$+(x-4)(x-5)(x-7)(1)+\ldots$$

Put $x=8$

$$f(8)=48+208+180+12$$

$$=448$$

6. Let us find the polynomial f(x) by using Lagrange's formula and hence find f(3) for the following values of x and y.

x_0	1	2	5
y_2	3	12	147

Solution:

Given: $x_0 = 0;\ x_1 = 1;\ x_2 = 2;\ x_3 = 5$

$$y_0 = 2;\ y_1 = 3;\ y_2 = 12;\ y_3 = 147$$

Lagrange's interpolation formula is:

$$y = f(x) = \frac{(x-x_1)(x-x_2)(x-x_3)}{(x_0-x_1)(x_0-x_2)(x_0-x_3)}y_0 + \frac{(x-x_0)(x-x_2)(x-x_3)}{(x_1-x_0)(x_1-x_2)(x_1-x_3)}y_1$$
$$+\frac{(x-x_0)(x-x_1)(x-x_3)}{(x_2-x_0)(x_2-x_1)(x_2-x_3)}y_2 + \frac{(x-x_0)(x-x_1)(x-x_2)}{(x_3-x_0)(x_3-x_1)(x_3-x_2)}y_3$$

$$f(x) = \frac{(x-1)(x-2)(x-5)}{(0-1)(0-2)(0-5)}(2) + \frac{(x-0)(x-2)(x-5)}{(1-0)(1-2)(1-5)}(3)$$
$$+\frac{(x-0)(x-1)(x-5)}{(2-0)(2-1)(2-5)}(12) + \frac{(x-0)(x-1)(x-2)}{(1-0)(1-2)(5-2)}(147)$$

$$f(x) = -\frac{1}{5}\left(x^3-8x^2+17x-10\right) + \frac{3}{4}\left(x^3-7x^2+10x\right)$$
$$-2\left(x^3-6x^2+5x\right) + \frac{49}{20}\left(x^3-3x^2+2x\right)$$

$$f(x) = x^3 + x^2 - x + 2 \qquad ...(1)$$

Put $x = 3$ in (1)

$$f(3) = 3^3 + 3^2 - 3 + 2$$

$$f(3) = 35$$

1.3.2 Central Difference Interpolation Formulae

Gauss Forward Central Difference Formula

If $\ldots, x_{-2}, x_{-1}, x_0, x_1, x_2, \ldots$ are given set of observation with common difference h and let $\ldots, y_{-2}$, are their correspond value where $y = f(x)$ be the given function then

$$y_p = y_0 + p\Delta y_{-1} + \frac{p(p+1)}{2!}\Delta^2 y_{-1} + \frac{p(p+1)(p-1)}{3!}\Delta^3 y_{-2} + \frac{p(p+1)(p-1)(p+2)}{4!}\Delta^4 y_{-2} + \ldots$$

Where, $p = \dfrac{x - x_0}{h}$

Proof:

x	y = f(x)	Δy	Δ²y	Δ³y	Δ⁴y
⋮	⋮				
x_{-2}	y_{-2}				
		Δy_{-2}			
x_{-1}	y_{-1}		$\Delta^2 y_{-2}$		
		Δy_{-1}		$\Delta^3 y_{-2}$	
x_0	y_0		$\Delta^2 y_{-1}$		$\Delta^4 y_{-2}$
x_1	y_1	Δy_0		$\Delta^3 y_{-1}$	
			$\Delta^2 y_0$		
x_2	y_2				
		Δy_{-1}			
⋮	⋮				

Let us assume a polynomial equation by using the arrow marks shown in the above table:

$$\text{Let, } y_p = y_0 + G_1 \Delta y_0 + G_2 \Delta^2 y_{-1} + G_3 \Delta^3 y_{-1} + G_4 \Delta^4 y_{-2} + \ldots \qquad \ldots(1)$$

Where $G_0, G_1, G_2, \ldots$ are unknown

$$y_p = y_{p+0} = E^p y_0 = (1+\Delta)^p y_0 \quad (\because E = 1+\Delta)$$

$$\Rightarrow y_p = (1 + pc_1\Delta + pc_2\Delta^2 + pc_3\Delta^3 + \ldots + pc_p\Delta^p) y_0$$

$$\Rightarrow y_p = y_0 + p\Delta y_0 + \frac{p(p-1)}{2!}\Delta^2 y_0 + \frac{p(p-1)(p-2)}{3!}\Delta^3 y_0 + \ldots \qquad \ldots(2)$$

Now, $y_{-1} = y_{-1+0} = E^{-1} y_0 = (1+\Delta)^{-1} y_0$

$$= (1 - \Delta + \Delta^2 - \Delta^3 + \ldots) y_0$$

$$\Rightarrow y_{-1} = y_0 - \Delta y_0 + \Delta^2 y_0 - \ldots$$

Therefore, $\Delta^2 y_{-1} = \Delta^2 y_0 - \Delta^3 y_0 + \ldots$ ...(3)

And $\Delta^3 y_{-1} = \Delta^3 y_0 - \Delta^4 y_0 + \ldots$ ...(4)

Substituting 2, 3, 4 in 1, we get:

$$y_0 + p\Delta y_0 + \frac{p(p-1)}{2!}\Delta^2 y_0 + \frac{p(p-1)(p-2)}{3!}\Delta^3 y_0 + \ldots = y_0 + G_1 \Delta y_0 + G_2\left(\Delta^2 y_0 - \Delta^3 y_0 + \cdots\right) + G_3(\Delta^3 y_0 - \Delta^4 y_0 + \ldots) + \ldots$$

Comparing corresponding coefficients, we get:

$$G_1 = p, G_2 = \frac{p(p-1)}{2!}, -G_2 + G_3 = \frac{p(p-1)(p-2)}{3!} \Rightarrow G_3 = \frac{p(p-1)(p+1)}{3!}$$

Similarly, $G_4 = \dfrac{p(p-1)(p+1)(p-2)}{4!}$

Substituting all these values of $G_0, G_1, G_2, \ldots$ in (1), we get:

$$y_p = y_0 + p\Delta y_0 + \frac{p(p-1)}{2!}\Delta^2 y_{-1} + \frac{p(p-1)(p+1)}{3!}\Delta^3 y_{-1} + \frac{p(p-1)(p+1)(p-2)}{4!}\Delta^4 y_{-2} + \ldots$$

Gauss Backward Central Difference Formula

If $\ldots, x_{-2}, x_{-1}, x_0, x_1, x_2, \ldots$ are given set of observation with common difference h and let $\ldots, y_{-2},$ are their corresponding values where $y = f(x)$ be the given function then:

$$y_p = y_0 + p\Delta y_{-1} + \frac{p(p+1)}{2!}\Delta^2 y_{-1} + \frac{p(p+1)(p-1)}{3!}\Delta^3 y_{-2} + \frac{p(p+1)(p-1)(p+2)}{4!}\Delta^4 y_{-2} + \ldots$$

Where $p = \dfrac{x - x_0}{h}$

Proof:

x	y	Δy	$\Delta^2 y$	$\Delta^3 y$	$\Delta^4 y$
⋮	⋮				
x_{-2}	y_{-2}				
		Δy_{-2}			
x_{-1}	y_{-1}		$\Delta^2 y_{-2}$		
		Δy_{-1}		$\Delta^3 y_{-2}$	

x_0	y_0		$\Delta^2 y_{-1}$		$\Delta^4 y_{-2}$
x_1	y_1	Δy_0		$\Delta^3 y_{-1}$	
			$\Delta^2 y_0$		
x_2	y_2				
		Δy_{-1}			
⋮	⋮				

Let us assume a polynomial equation by using the arrow marks shown in the above table.

$$\text{Let } y_p = y_0 + G_1\Delta y_{-1} + G_2\Delta^2 y_{-1} + G_3\Delta^3 y_{-2} + G_4\Delta^4 y_{-2} + \dots \qquad \text{...(1)}$$

Where $G_0, G_1, G_2, \dots$ are unknowns:

$$y_p = y_{p+0} = E^p y_0 = (1+\Delta)^p y_0 \quad (\because E = 1+\Delta)$$

$$\Rightarrow y_p = (1 + pc_1\Delta + pc_2\Delta^2 + pc_3\Delta^3 + \dots + pc_p\Delta^p)y_0$$

$$\Rightarrow y_p = y_0 + p\Delta y_0 + \frac{p(p-1)}{2!}\Delta^2 y_0 + \frac{p(p-1)(p-2)}{3!}\Delta^3 y_0 + \dots \qquad \text{...(2)}$$

Now, $y_{-1} = y_{-1+0} + E^{-1} y_0 = (1+\Delta)^{-1} y_0$

$$= (1 - \Delta + \Delta^2 - \Delta^3 + \dots)y_0$$

$$\Rightarrow y_{-1} = y_0 - \Delta y_0 + \Delta^2 y_0 - \dots$$

Therefore, $\Delta y_{-1} = \Delta y_0 - \Delta^2 y_0 + \Delta^3 y_0 \dots$ ...(3)

$$\Delta^2 y_{-1} = \Delta^2 y_0 - \Delta^3 y_0 + \dots \qquad \text{...(4)}$$

Also $y_{-2} = y_{-2+0} = E^{-2} y_0 = (1+\Delta)^{-2} y_0$

$$= (1 - 2\Delta + 2\Delta^2 - 4\Delta^3 + \dots)y_0$$

$$\Rightarrow y_{-2} = y_0 - 2\Delta y_0 + 3\Delta^2 y_0 - \cdots$$

Now, $\Delta^3 y_{-2} = \Delta^3 y_0 - 2\Delta^4 y_0 + \dots$ ...(5)

Substituting 2, 3, 4, 5 in 1, we get:

$$y_0+p\Delta y_0+\frac{p(p-1)}{2!}\Delta^2 y_0+\frac{p(p-1)(p-2)}{3!}\Delta^3 y_0+\ldots=y_0+G_1\left(\Delta y_0-\Delta^2 y_0+\Delta^3 y_0-\ldots\right)+$$

$$G_3(\Delta^2 y_0-\Delta^3 y_0+\Delta^4 y_0-\ldots)+G_3\left(\Delta^3 y_0-2\Delta^4 y_0+\ldots\right)+\ldots$$

Comparing corresponding coefficients, we get:

$$G_1=p, G_1+G_2=\frac{p(p-1)}{2!}\Rightarrow G_2=\frac{p(p+1)(p-2)}{2!}$$

Also, $G_1-G_2+G_3=\frac{p(p-1)(p-2)}{3!}\Rightarrow G_3=\frac{p(p+1)(p-1)}{3!}$

Similarly, $G_4=\frac{p(p+1)(p-1)(p+2)}{4!},\ldots$

Substituting all these values of $G_0, G_1, G_2, \ldots$ in (1), we get:

$$y_p=y_0+p\Delta y_{-1}+\frac{p(p+1)}{2!}\Delta^2 y_{-1}+\frac{p(p+1)(p-1)}{3!}\Delta^3 y_{-2}$$

$$+\frac{p(p+1)(p-1)(p+2)}{4!}\Delta^4 y_{-2}+\ldots, p=\frac{x-x_0}{h}$$

Stirling's Formulae

If $\ldots, x_{-2}, x_{-1}, x_0, x_1, x_2, \ldots$ are given set of observation with common difference h and let $\ldots, y_{-2}, y_{-1}, y_0, y_1, y_2, \ldots$ are their corresponding values, where $y=f(x)$ be the given function then:

$$y_p=y_0+p\left(\frac{\Delta y_0+\Delta y_{-1}}{2}\right)+\frac{p^2}{2!}\Delta^2 y_{-1}+\frac{p(p^2-1)}{3!}\left(\frac{\Delta^2 y_{-1}+\Delta^2 y_{-2}}{2}\right)+\frac{p^2(p^2-1)}{4!}\Delta^4 y_{-2}\ldots$$

Where $p=\frac{x-x_0}{h}$

Stirling's formula will be obtained by taking the average of Gauss forward difference formula and gauss backward difference formula.

We know Gauss backward difference formula:

$$y_p=y_0+p\Delta y_0+\frac{p(p-1)}{2!}\Delta^2 y-1+\frac{p(p-1)(p+1)}{3!}\Delta^3 y_{-2}$$

$$+\frac{p(p-1)(p+1)(p-2)}{4!}\Delta^4 y_{-2}\cdots>(1)$$

Also, from Gauss backward difference formula:

$$y_p = y_0 + p\Delta y_{-1} + \frac{p(p+1)}{2!}\Delta^2 y_{-1} + \frac{p(p-1)(p+1)}{3!}\Delta^3 y_{-2}$$

$$+\frac{p(p+1)(p-1)(p+2)}{4!}\Delta^4 y_{-2} + \cdots > (2)$$

Now stirling's Formula $=\frac{1}{2}$ (Gauss forward formula + Gauss backward formula):

$$\therefore y_p = y_0 + p\left(\frac{\Delta y_0 + \Delta y_{-1}}{2}\right) + \frac{p^2}{2!}\Delta^2 y_{-1}$$

$$+\frac{p(p^2-1)}{3!}\left(\frac{\Delta^3 y_{-1} + \Delta^3 y_{-2}}{2}\right) + \frac{p^2(p^2-1)}{4!}\Delta^4 y_{-2} + \ldots$$

1.4 Interpolation with Unevenly Spaced Points-Lagrange's Interpolation Formula

Interpolation with Unequal Intervals

Interpolation is a process of estimating the value of a function at an intermediate point when its value is known only at certain specified points. It is based on the following assumptions:

- Given equation is a polynomial or it can be represented by a polynomial with a good degree of approximation.
- Function should vary in such a way that either it its increasing or decreasing in the given range without sudden jumps or falls of functional values in the given interval.

We shall discuss the concept of interpolation from a set of tabulated values of y = f(x) when the values of x are given intervals or at unequal intervals. First we consider interpolation with unequal intervals.

Lagrange's interpolation formula for unequal intervals:

Let y = f(x) be a function which takes the values $y_0, y_1 y_2 \ldots y_n$ corresponding to $x_0, x_1, x_2, \ldots, x_n$.

Then Lagrange's interpolation formula is:

$$y = f(x)\frac{(x-x_1)(x-x_2)\ldots(x-x_n)}{(x_0-x_1)(x_0-x_2)\ldots(x_0-x_n)}y_0 + \frac{(x-x_0)(x-x_2)\ldots(x-x_n)}{(x_1-x_0)(x_1-x_2)\ldots(x_1-x_n)}y_1 +$$

$$\ldots\frac{(x-x_0)(x-x_1)\ldots(x-x_{n-1})}{(x_n-x_0)(x_n-x_1)\ldots(x_n-x_{n-1})}y_n$$

Problems

1. Let us find the second degree polynomial through the points (0, 2), (2,1), (1,0) using Lagrange's formula.

Solution:

Given: $x_0 = 0;\ x_1 = 2;\ x_2 = 1$

$y_0 = 2;\ y_1 = 1;\ y_2 = 0$

Lagrange's formula is:

$$y(x) = \frac{(x-x_1)(x-x_2)}{(x_0-x_1)(x_0-x_2)}y_0 + \frac{(x-x_0)(x-x_2)}{(x_1-x_0)(x_1-x_2)}y_1 + \frac{(x-x_0)(x-x_1)}{(x_2-x_0)(x_2-x_1)}y_2$$

$$= \frac{(x-2)(x-1)}{(-2)(-1)}(2) + \frac{(x-0)(x-1)}{(2)(1)}(1) + 0$$

$$= x^2 - 3x + 2 + \frac{1}{2}(x^2 - x)$$

$$y(x) = \frac{3x^2}{2} - \frac{7x}{2} + 2$$

2. Using Lagrange's formula, let us find the polynomial to the given data.

X:	0	1	3
Y:	5	6	50

Solution:

Given:

$x_0 = 0;\ x_1 1;\ x_2 = 3$

$y_0 = 5;\ y_1 = 6; y_2 = 50$

Lagrange's formula is:

$$y=f(x)=\frac{(x-x_1)(x-x_2)}{(x_0-x_1)(x_0-x_2)}y_0+\frac{(x-x_0)(x-x_2)}{(x_1-x_0)(x_1-x_2)}y_1+\frac{(x-x_0)(x-x_1)}{(x_2-x_0)(x_2-x_1)}y_2$$

$$=\frac{(x-1)(x-3)(5)}{(-1)(-3)}+\frac{x(x-3)(6)}{(-1)(-2)}+\frac{x(x-1)(50)}{(3)(2)}$$

$$=\frac{5}{3}(x^2-4x+3)-3x^2+9x+\frac{25}{3}x^2-\frac{25x}{3}$$

$$f(x)=7x^2-6x+5$$

3. Apply Lagrange's formula, to find y(27) to the data given below.

x:	14	17	31	35
y:	68.8	64	44	39.1

Given:

$$x_0=14;\ x_1=17;=31 \text{ and } x_3=35$$

$$y_0=68.8; y_1=64;\ y_2=44 \text{ and } y_3=39.1$$

Solution:

$$y(x)=\frac{(x-x_1)(x-x_1)(x-x_3)}{(x_0-x_1)(x-x_1)(x_0-x_3)}y_0+\frac{(x-x_0)(x-x_2)(x-x_3)}{(x_1-x_0)(x_1-x_2)(x_1-x_3)}y_1$$

$$+\frac{(x-x_0)(x-x_1)(x-x_3)}{(x_2-x_0)(x_2-x_1)(x_2-x_3)}y_2+\frac{(x-x_0)(x-x_1)(x-x_2)}{(x_3-x_0)(x_3-x_1)(x_3-x_2)}y_3$$

Put 27

$$y(27)=\frac{(27-17)(27-31)(27-35)}{(14-17)(14-31)(14-35)}(68.8)$$

$$+\frac{(27-14)(27-31)(27-35)}{(17-14)(27-31)(27-35)}(64)$$

$$+\frac{(27+14)(27-17)(27-35)}{(31-14)(31-17)(31-35)}(44)$$

$$+\frac{(27-14)(27-17)(27-31)}{(35-14)(35-17)(35-31)}(39.1)$$

$$y(27)=-20.556+35.217+48.067-13.447$$

$$y(27)=49.281$$

4. Using Lagrange's interpolation formula, let us find y(2) from the following data: y(0)=0; y(l)=1; y(3)=81; y(4)=256; y(5)=625.

Solution:

Given:

X	x_0	x_1	x_2	x_3	x_4
	0	1	3	4	5
y(x)	0	1	81	256	625
	y_0	y_1	y_2	y_3	y_4

Lagrange interpolation formula is:

$$y=f(x)=\frac{(x-x_1)(x-x_2)(x-x_3)(x-x_4)}{(x_0-x_1)(x_0-x_2)(x_0-x_3)(x_0-x_4)}y_0$$

$$+\frac{(x-x_0)(x-x_2)(x-x_3)(x-x_4)}{(x_1-x_0)(x_1-x_2)(x_1-x_3)(x_1-x_4)}y_1$$

$$+\frac{(x-x_0)(x-x_1)(x-x_3)(x-x_4)}{(x_2-x_0)(x_2-x_1)(x_2-x_3)(x_2-x_4)}y_2$$

$$+\frac{(x-x_0)(x-x_1)(x-x_2)(x-x_4)}{(x_3-x_0)(x_3-x_1)(x_3-x_2)(x_3-x_4)}y_3$$

$$+\frac{(x-x_0)(x-x_1)(x-x_2)(x-x_3)}{(x_4-x_0)(x_4-x_1)(x_4-x_2)(x_4-x_3)}y_4$$

$$y(2)=\frac{(2-1)(2-3)(2-4)(2-5)}{(0-1)(0-3)(0-4)(0-5)}(0)$$

$$+\frac{(2-0)(2-3)(2-4)(2-5)}{(1-0)(1-3)(1-4)(1-5)}(1)$$

$$+\frac{(2-0)(2-1)(2-4)(2-5)}{(3-0)(3-1)(3-4)(3-5)}(81)$$

$$+\frac{(2-0)(2-1)(2-3)(2-5)}{(4-0)(4-1)(4-3)(4-5)}(256)$$

$$+\frac{(2-0)(2-1)(2-3)(2-4)}{(5-0)(5-1)(5-3)(5-4)}(625)$$

$$y(2)=0+\frac{12}{24}+\frac{(12)(81)}{12}-\frac{(6)(256)}{12}+\frac{(4)(625)}{40}$$

$$=16$$

5. Use Lagrange's interpolation formula to fit a polynomial to the given data f(- 1) = - 8, f(0) = 3, f(2) = 1 and f(3) = 2. Let us find the value of f(1).

Solution:

Given: $x_0 = -1;\ x_1 = 0;\ x_2 = 2;\ x_3 = 3,$

$$y_0 = -8;\ y_0 = 3;\ y_2 = 1;\ y_3 = 2$$

Lagrange's interpolation formula is:

$$y=f(x)=\frac{(x-x_1)(x-x_2)(x-x_3)y_0}{(x_0-x_1)(x_0-x_2)(x_0-x_1)}+\frac{(x-x_0)(x-x_2)(x-x_3)y_1}{(x_1-x_0)(x_1-x_2)(x_1-x_3)}$$

$$+\frac{(x-x_0)(x-x_1)(x-x_3)y_2}{(x_2-x_0)(x_2-x_1)(x_2-x_3)}+\frac{(x-x_0)(x-x_1)(x-x_2)y_3}{(x_3-x_0)(x_3-x_1)(x_3-x_2)}$$

$$y=\frac{x(x-2)(x-3)(-8)}{(-1)(-3)(-4)}+\frac{(x+1)(x-2)(x-3)(3)}{(1)(-2)(-3)}$$

$$+\frac{(x+1)(x)(x-3)(1)}{(3)(2)(-1)}+\frac{(x+1)(x)(x-2)(2)}{(4)(3)(1)}$$

$$=\frac{2}{3}(x^3-5x^2+6x)+\frac{1}{2}(x^3-4x^2+x+6)$$

$$-\frac{1}{6}(x^3-2x^2-3x)+\frac{1}{6}(x^3-x^2-2x)$$

$$f(x) = \frac{7}{6}x^3 - \frac{31}{6}x^2 + \frac{14}{3}x + 3$$

$$f(1) = \frac{7}{6} - \frac{31}{6} + \frac{14}{3} + 3$$

$$f(1) = \frac{11}{3}$$

6. Using Lagrange's method, let us find log10 656, given that log10 654=2.8156, log10 658=2.8182, log10 659=2.8189 and log10 661 = 2.8202.

Solution:

Given:

	x_0	x_1	x_2	x_3
x	654	658	659	661
$y = \log_{10} x$	2.8156	2.8182	2.8189	2.8202
	y_0	y_1	y_2	y_3

To find y, when x = 656.

Let $y = \log_{10}^{x}$

Lagrange's formula is:

$$y(x) = \frac{x(x-x_1)(x-x_2)(x-x_3)}{(x_0-x_1)(x_0-x_2)(x_0-x_3)}y_0$$

$$+\frac{(x-x_0)(x-x_2)(x-x_3)}{(x_1-x_0)(x_1-x_2)(x_1-x_3)}y_1$$

$$+\frac{(x-x_0)(x-x_1)(x-x_3)}{(x_2-x_0)(x_2-x_1)(x_2-x_3)}y_2$$

$$+\frac{(x-x_0)(x-x_1)(x-x_2)}{(x_3-x_0)(x_3-x_1)(x_3-x_2)}y_3$$

$$y(656) = \frac{(656-658)(656-659)(656-661)(2.8156)}{(654-658)(654-659)(654-661)}$$

$$+\frac{(656-654)(656-659)(656-661)(2.8182)}{(658-654)(658-659)(658-661)}+$$

$$\frac{(656-654)(656-658)(656-661)(2.8189)}{(659-654)(659-658)(659-661)}$$

$$+\frac{(656-654)(656-658)(656-659)(2.8202)}{(661-654)(661-658)(661-659)}$$

i.e., $\log_{10} 656 = 0.6033+7.0455-5.6378+0.8058$

7. Using Lagrange's formula, let us find the value of y at x = 6 from the following data:

x:	3	7	9	10
y:	168	120	72	63

Solution:

Given:

$x_0 = 3$;	$x_1 = 7$;	$x_2 = 9$;	$x_3 = 10$.
$y_0 = 168$;	$y_1 = 120$;	$y_2 = 72$;	$y_3 = 63$.

Lagrange's formula is,

$$y(x)=\frac{(x-x_1)(x-x_2)(x-x_3)}{(x_0-x_1)(x_0-x_2)(x_0-x_3)}y_0$$

$$+\frac{(x-x_0)(x-x_2)(x-x_3)}{(x_1-x_0)(x_2-x_1)(x_2-x_3)}y_1$$

$$+\frac{(x-x_0)(x-x_1)(x-x_3)}{(x_2-x_0)(x_2-x_1)(x_2-x_3)}y_2$$

$$+\frac{(x-x_0)(x-x_1)(x-x_2)}{(x_3-x_0)(x_3-x_1)(x_3-x_2)}y_3 \qquad ...(1)$$

To find y, at x = 6 i.e., y (6).

Put x = 6 and given data in (1).

$$y(6)=\frac{(6-7)(6-9)(6-10)(168)}{(3-7)(3-9)(3-10)}+\frac{(6-3)(6-9)(6-10)(120)}{(7-3)(7-9)(7-10)}$$

$$=\frac{(6-3)(6-7)(6-10)(72)}{(9-3)(9-7)(9-10)}+\frac{(6-3)(6-7)(6-9)(63)}{(10-3)(10-7)(10-9)}$$

$$=\frac{(-1)(-3)(-4)(168)}{(-4)(-6)(-7)}+\frac{(-3)(-3)(-4)(120)}{(-4)(-2)(-3)}$$

$$+\frac{(-3)(-1)(-4)(72)}{(6)(2)(-1)}+\frac{(3)(-1)(-3)(63)}{(7)(3)(1)}$$

$$=12+180-72+27$$

$$y(6)=147$$

8. Using Lagrange's interpolation, let us calculate the profit in the year 2000 from the following data:

Year	1997	1999	2001	2002
Profit in Lakhs of Rs.	43	65	159	248

Solution:

Given: $x_0=1997, x_1=1999;\ x_2=2001,\ x_3=2002$

$$y_0=43;\ y_1=65;\ y_2=159;\ y_3=248$$

Lagrange's Interpolation formula is:

$$y(x)=\frac{(x-x_1)(x-x_2)(x-x_3)}{(x_0-x_1)(x_0-x_2)(x_0-x_3)}y_0$$

$$+\frac{(x-x_0)(x-x_2)(x-x_3)}{(x_1-x_0)(x_1-x_2)(x_1-x_3)}y_1$$

$$+\frac{(x-x_0)(x-x_1)(x-x_3)}{(x_2-x_0)(x_2-x_1)(x_2-x_3)}y_2$$

$$+\frac{(x-x_0)(x-x_1)(x-x_2)}{(x_3-x_0)(x_3-x_1)(x_3-x_2)}y_3 \qquad ...(1)$$

Profit in the year x = 2000 is:

$$y(2000)=\frac{(2000-1999)(2000-2001)(2000-2002)}{(1997-1999)(1997-2001)(1997-2002)}43+$$

$$\frac{(2000-1997)(2000-2001)(2000-2002)}{(1999-1997)(1999-2001)(1999-2003)}65+$$

$$\frac{(2000-1997)(2000-1999)(2000-2002)}{(2001-1997)(2001-1999)(2001-2002)}159+$$

$$\frac{(2000-1997)(2000-1999)(2000-2001)}{(2002-1997)(2002-1999)(2002-2001)}248$$

$$y(2000)=\frac{(1)(-1)(-2)(43)}{(-2)(-4)(-5)}+\frac{(3)(-1)(-2)(65)}{(2)(-2)(-3)}$$

$$=+\frac{(3)(1)(-2)(159)}{(4)(2)(-1)}+\frac{(3)(1)(-1)(248)}{(5)(3)(1)}$$

$$=-\frac{86}{40}+\frac{139}{12}+\frac{954}{8}-\frac{744}{15}$$

$$=-2.15+32.5+119.25-49.6$$

$$y(2000)=100$$

The profit in the year 2000 is Rs. 100 lakhs.

1.4.1 B-Spline Interpolation

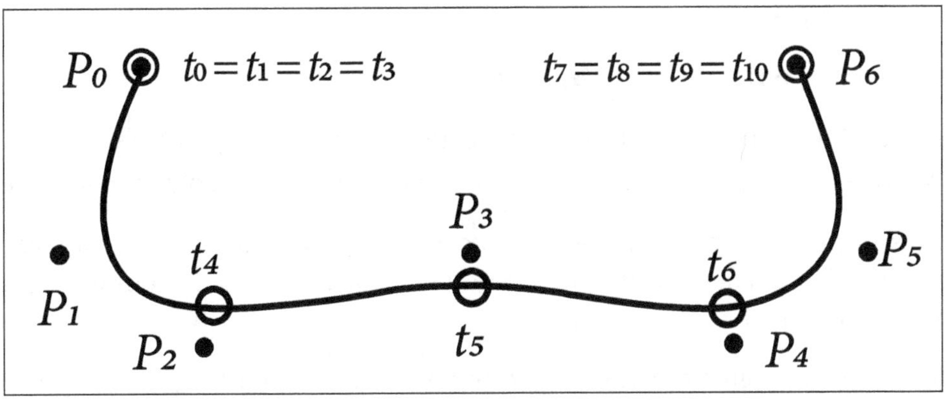

A B-spline is a generalization of the Bézier curve. Let a vector known as the knot vector be defined:

$$T=\{t_0,t_1,\ldots t_m\},$$

Where T is a non-decreasing sequence with $t_1 \in [0,1]$, and define control points $P_0,\ldots,P_n$ Define the degree as:

$$P \equiv m - n - 1$$

The "knots" $t_p + 1,\ldots, t_{m-p-1}$ are called internal knots.

Define the basic functions as:

$$= \begin{cases} 1 \ \text{if } t_1 \le t \le t_i < t_{i+1} \text{ and } t_i < t_i + 1 \\ 0 \ \text{otherwise} \end{cases}$$

$$= \frac{t - t_i}{t_{i+j} - t_i} N_{i.j-1}(t) + \frac{t_{i+j+1} - t}{t_{i+j+1} - t_{i+1}} N_{i+1.j-1}(t)$$

Where $j = 1, 2, \ldots, P$. Then the curve defined by:

$$C(t) = \sum_{i=0}^{n} P_i N_{i \cdot p}(t)$$

is B-spline.

Specific types include the non-periodic B-spline and uniform B-spline. A B-spline with no internal knots is a Bézier curve.

A curve is $p - k$ times differentiable at a point where k duplicate knot values occur. The knot values determine the extent of the control of the control points.

1.4.2 Cubic Spline

Interpolating with a Cubic Spline

We consider the problem of interpolation between given data points (x_i, y_i), i=0, 1, 2, 3....n where $a = x_1 < x_2 < x_3 < \ldots < x_n = -b$ by means of a smooth polynomial curve.

By means of method of least squares, we can fit a polynomial but it is appropriate. "Spline fitting" is the new technique recently developed to fit a smooth curve passing he given set of points.

Definition: A cubic spline function S(x) with respect to the points $x_0, x_1, \ldots, x_n$ is a polynomial of degree three in each interval $(x_i - t, x_i)$; i = 1, 2, ..., n. such that S(x), S (x) and S(x) are continuous in (x_0, x_n).

Conditions for Fitting Spline Fit

The conditions for a cubic spline fit are that we pass a set of cubic through the points, using a new cubic in each interval. Further it is required both the slope and the curvature be the same for the pair of cubic that join at each point.

Problems

1. Fit the cubic splines for the following data:

x:	1	2	3	4	5
y:	1	0	1	0	1

Solution:

Given: $x_0 = 1,\ x_1 = 2,\ x_2 = 3,\ x_3 = 4,\ x_4 = 5$

$y_0 = 1.\ y_1 = 0,\ y_2 = 1,\ y_3 = 0,\ y_4 = 1$

Let us assume, $M_0 = M_4 = 0$

Here h = 1 and n = 4

We have:

$$M_{i-1} + 4M_i + M_{i+1} = \frac{6}{h^2}\left[y_i - 2y_1 + y_{i+1}\right]$$

For $i = 1,2,3$ ...(1)

$$\therefore\ M_0 + 4\,M_1 + M_2 = 6\left[y_0 - 2y_1 + y_2\right]$$

$$M_1 + 4\,M2 + M3 = 6\left[y_1 - 2y_2 + y_3\right]$$

$$M_2 + 4\,M_3 + M_4 = 6\left[y_2 - 2y_3 + y_4\right]$$

i.e., $4\,M_1 + M_2 = 6\left[1 - 0 + 1\right]$

$4M_1 + M_2 = 12$ ...(2)

$M_1 + 4\,M_2 + M_3 = 6\left[0 - 2 + 0\right]$

i.e., $M_1 + 4\,M_2 + M_3 = -12$ ...(3)

$M_2 + 4\,M_3 + M_4 = 6\left[1 - 0 + 1\right]$

i.e., $M_2 + 4\,M_3 = 12$ ...(4)

Solving (2), (3) and (4) we get:

$$M_1 = 4.286;\ M_2 = -5.143\ \&\ M_3 = 4.286$$

The equation for Cubic Spline is:

$$S(x) = \frac{1}{6h}\left[(x_i - x)^3 M_{i-1} + (x - x_i - 1)^3 M_i\right]$$

$$+\frac{1}{h}(x_i - x)\left[y_{i-1} - \frac{h^2}{6}M_{i-1}\right]$$

$$+\frac{1}{h}(x - x_{i-1})\left[y_i - \frac{h^2}{6}M_i\right] \qquad ...(2)$$

$$i = 1,2,3,4.$$

Case (i): Cubic Spline corresponding to the interval [1 2], put i = 1 in (2).

$$S(x) = \frac{1}{6}\left[(x_i - x)^3 M_0 + (x - x_0)^3 M_1\right] + (x_1 - x)\left[y_0 - \frac{1}{6}M_0\right]$$

$$+(x - x_0)\left[y_1 - \frac{1}{6}M_1\right]$$

$$= \frac{1}{6}\left[4.286(x-1)^3\right] + 2(2-x)(1) + (x-1)\left[0 - \frac{4.286}{6}\right]$$

$$S(x) = 0.7143(x-1)^3 + 2(2-x) - 0.7143(x-1) \qquad ...(3)$$

Case (ii): Cubic Spline Corresponding to the interval [2, 3] put i = 2 in (2).

$$S(x) = \frac{1}{6}\left[(x_2 - x)^3 M_1 + (x - x_1)^3 M_2\right] + (x_2 - x)\left[y_1 - \frac{1}{6}M_1\right]$$

$$+(x - x_1)\left[y_2 - \frac{1}{6}M_2\right]$$

$$= \frac{1}{6}\left[(3-x)^3(4.286) - 5.143(x-2)^3\right]$$

$$+(3-x)\left[0 - \frac{4.286}{6}\right] + (x-2)\left[1 + \frac{5.143}{6}\right]$$

$$S(x)=0.7143(3-x)^3-0.8572(x-2)^3$$

$$-0.7143(3-x)+1.8572(x-2) \qquad ...(4)$$

Case (iii): Cubic Spline Corresponding to the interval [3, 4] put i = 3 in (2).

$$S(x)=\frac{1}{6}\left[(x_3-x)^3 M_2+(x-x_2)^3 M_3\right]+(x_3-x)\left[y_2-\frac{1}{6}M_2\right]$$

$$+(x-x_2)\left[y_3-\frac{1}{6}M_3\right]$$

$$=\frac{1}{6}\left[-5.143(4-x)^3+4.286(x-3)^3\right]+(4-x)\left[1+\frac{5.143}{6}\right]$$

$$+(x-3)\left[0-\frac{4.286}{6}\right]$$

$$S(x)=0.8572(4-x)^3-0.7143(x-3)^3$$

$$+1.8572(4-x)-0.7143(x-3) \qquad ...(5)$$

Case (iv): Cubic Spline Corresponding to the interval [4, 5] put i = 4 in (2).

$$S(x)=\frac{1}{6}\left[(x_4-x)^3 M_3+(x-x_3)^3 M_4\right]+(x_4-x)\left[y_3-\frac{1}{6}M_3\right]$$

$$+(x-x_3)\left[y_4-\frac{1}{6}M_4\right]$$

$$S(x)=0.7143(5-x)^3-0.7143(5-x)+(x-4) \qquad ...(6)$$

2. Let us find the natural cubic spline approximation for the function f(x) defined by the following data:

x:	0	1	2	3
f(x):	1	2	33	244

Also find the value of f(2.5) and f'(2.5)

Given: $x_0 = 0$; $x_1 = 1$; $x_2 = 2$ and $x_3 = 3$

$$y_0 = 1; y_1 = 2; y_0 = 33 \text{ and } y_3 = 244.$$

Solution:

Since x are equally spaced with h = 1 and n = 4.

$$M_i - 1 + 4M_i + M_i + 1 = 6/h^2 (y_{i-1} - 2y_i + y_{i+1}) \quad ...(1)$$

Where i = 1,2,3, and let $M_0 = 0$; $M_3 = 0$.

Put i = 1 in (1):

$$M_0 + 4M_1 + M_2 = 6(y_0 - 2y_1 + y_2)$$

$$4M_1 + M_2 = 6(1 - 4 + 33) = 180 \text{ (i)}$$

Put i = 2 in (1):

$$M_1 + 4M_2 + M_3 = 6(y_1 - 2y_2 + y_3)$$

$$M_1 + 4M_2 = 6(2 - 66 + 244) = 1080 \text{ (ii)}$$

Solving (i) and (ii) we get $M_1 = -24, M_2 = 276$

Now the required cubic spline is:

$$y(x) = \frac{1}{6h}\left[(x_i - x)^3 M_{i-1} + (x - x_{i-1})^3 M_i\right]$$

$$+ \frac{1}{h}(x_i - x)\left(y_{i-1} - \frac{h^2}{6} M_{i-1}\right)$$

$$+ \frac{(x - x_{i-1})}{h}\left(y_i - \frac{h^2}{6} M_i\right) ...(1)$$

For $0 \le x \le 1$ put i = 1 in (1):

$$y(x) = \frac{1}{6}\left[(x_1 - x)^3 M_0 + (x - x_0)^3 M_1\right] + (x - x)$$

$$\left(y_0 - \frac{M_0}{6}\right) + (x - x_0)\left(y_1 - \frac{M_1}{6}\right)$$

$$= \frac{1}{6}(x)(-24) + (1 - x)(1) + (x)(2 + 4)$$

$$y(x) = 1 + 5x - 4x^3 \text{ in } (0,1)$$

For $1 \le x \le 2$ put i = 2 in (1):

$$y = \frac{1}{6}\left[(x_2 - x)^3 M_1 + (x - x_1)^3 M_2\right] + (x_2 - x)$$

$$\left(y_1 - \frac{M_1}{6}\right) + (x - x_1)\left(y_1 - \frac{M_2}{6}\right)$$

$$= \frac{1}{6}\left[(2-x)^3(-24) + (x-1)^3(276)\right]$$

$$+(2-x)(3+4) + (x-1)(33-46)$$

$$f(x) = -4(2-x)^3 + 46(x-1)^3 + 6(2-x) - 13(x-1)(33-46) \text{ in } (1,2)$$

For $2 \le x \le 3$ put $i = 3$ in (1):

$$f(x) = \frac{1}{6}\left[(x_3 - x)^3 M_2 + (x - x_2)^3 M_3\right] + (x_3 - x)$$

$$\left(y_2 - \frac{M_2}{6}\right) + (x - x_2)\left(y_3 - \frac{M_3}{6}\right)$$

$$= \frac{1}{6}(3-x)^3(276) + (3-x)(33-46) + (x-2)(244)$$

$$f(x) = 46(3-x)^3 - 13(3-x) + 244(x-2) \text{ in}(2,3)$$

Now $f(x) = 138(3-x)^2 + 257$

$$\therefore f(2.5) = 46(3-2.5)^3 - 13(3-2.5) + 244(2.5-2)$$

$$= 121.25$$

$$f(2.5) = 138(3-2.5)^2 + 257$$

$$= 291.5.$$

3. Obtain the cubic spline for the following data to find y(0.5).

x:	-1	0	1	2
y:	-1	1	3	35

Solution:

Here x is equally spaced h = 1.

The cubic spline, in $x_i - 1 \le x \le x_i$; is given by:

$$y(x)=\frac{1}{6}\left[(x_i-x)^3 M_{i-1}+(x-x_{i-1})^3 M_i\right]$$

$$+(x_i-x)\left[y_{i-1}-\frac{M_{i-1}}{6}\right]+x-x_{i-1}\left[y_i-\frac{M_i}{6}\right] \quad ...(1)$$

Given:

$$x_0=-1;\ x_1=0;\ x_2=1;\ x_3=2,$$

$$y_0=-1;\ y_1=1;\ y_2=3;\ y_3=35\ \&\ M_0=0;\ M_3=0$$

$$M_i-1+4M_i+M_i+1=6(y_{i-1}-2y_i+y_i+1) \quad ...(2)$$

For i = 1,2.

Put i = 1, in (2) we get:

$$M_0+4M_1+M_2=0 \Rightarrow 4M_1+M_2=0 \quad ...(3)$$

Put i = 2, in (2) we get:

$$M_1+4M_2+M_3=6(y_1-2y_2+y_3)$$

i.e., $M_1+4M_2=180$ ...(4)

Solving (3) & (4) we get $M_1=-12,\ M_2=48$

(i) Put i = 1 in (1) we get polynomial in (-1, 0):

$$y(x)=\frac{1}{6}\left[(x_1-x)^3 M_0+(x-x_0)^3 M_1\right]$$

$$+(x_1-x)\left(y_0-\frac{M_0}{6}\right)+(x-x_0)\left[y_1-\frac{M_1}{6}\right] \quad ...(5)$$

$$y(x)=\frac{1}{6}\left[(x+1)^3(-12)\right]+(-x)(-1)+(x+1)\left(+1+\frac{12}{6}\right)$$

$$y(x)=-2x^3-6x^2-2x+1\ \text{in}(-1,0) \quad ...(A)$$

(ii) put i = 2 in (2) we get polynomial in (0,1):

$$y(x)=\frac{1}{6}\left[(x_2-x)^3M_1+(x-x_1)^3M_2\right]+(x_2-x)$$

$$\left(y_1-\frac{M_1}{6}\right)+(x-x_1)\left[y_2-\frac{M_2}{6}\right] \qquad ...(6)$$

$$y(x)=\frac{1}{6}\left[(1-x)^3(-12)+(x-0)^3(48)\right]+(1-x)$$

$$\left(1+\frac{12}{6}\right)+(x-0)\left(3-\frac{48}{6}\right)$$

$$y(x)=10x^2-6x^2-2x+1 \text{ in } (0,1) \qquad ...(B)$$

(iii) Put i = 3 in (1) we get polynomial in (1, 2):

$$y(x)=\frac{1}{6}\left[(x_3-x)^3M_2+(x-x_2)^3M_3\right]+(x_3-x)$$

$$\left(y_2-\frac{M_2}{6}\right)+(x-x_2)\left[y_3-\frac{M_3}{6}\right] \qquad ...(7)$$

$$y(x)=\frac{1}{6}\left[(2-x)^3(48)+(x-1)^3(0)\right]$$

$$+(2-x)\left(3-\frac{48}{6}\right)+(x-1)(35-0)$$

$$y(x)=8x^3-48x^2-56x+19 \text{ in } (-1,2) \qquad ...(C)$$

To find $y(0,5)$:

Put $x=0.5$ in (B) $\because 0.5$ lies in $(0,1)$

$$y(0,5)=10(0.5)^3-6(0.5)^2-2(0.5)+1$$

$$=0.25$$

4. Let us find the natural cubic spline to fit the data:

x:	0	1	2
f(x):	− 1	3	29

Hence find f (0.5) and f(1.5).

Solution:

Given:

$$x_0 = 0;\ x_1 = 1;\ x_2 = 2;\ y_0 - 1;\ y_1 = 3;\ y_2 = 29$$

Here h = 1 and n = 2

Also assume $M_0 = 0$ and $M_2 = 0$

We have:

$$M_{i-1} + 4M_i + M_{i+1} = \frac{6}{h^2}\left[y_{i-1} - 2y_i + y_{i+1}\right]$$

for $i = 1, 2, \ldots, (n-1)$

Put i = 1

$$M_0 + 4M_1 + M_2 = 6\left[y_0 - 2y_1 + y_2\right]$$

$$4M_1 = 6\left[-1 - 6 + 29\right]$$

$$M_1 = 33$$

$$S(x) = \frac{1}{6h}\left[(x_i - x)^3 M_{i-1} + (x - x_{i-1})^3 M_i\right]$$

$$+\frac{1}{h}(x_i - x)\left[y_{i-1} - \frac{h^2}{6}M_{i-1}\right]$$

$$+\frac{(x - x_{i-1})}{h}\left[y_i - \frac{h^2}{6}M_i\right] \quad \ldots(1)$$

For $0 \le x \le 1$ put i = 1 in (1):

$$S(x) = \frac{1}{6h}\left[(x_1 - x)^3 M_0 + (x - x_0)^3 M_1\right]$$

$$+\frac{(x_1 - x)}{h}\left[y_0 - \frac{M_0}{6}\right] + \frac{(x - x_0)}{h}\left(y_1 - \frac{M_1}{6}\right)$$

$$= \frac{1}{6}\left[33x^3\right] + (1 - x)(-1) + (x - 0)\left(3 - \frac{33}{6}\right)$$

$$= \frac{11}{2}x^3 - 1 + x - \frac{5x}{2}$$

$$f(x) = S(x) = \frac{11}{2}x^3 - \frac{3x}{2} - 1 \text{ in } (0 \le x \le 1) \qquad \ldots(2)$$

For $1 < x \le 2$. Put i = 2 in (1).

$$S(x) = \frac{1}{6}\left[(x_2 - x)^3 M_1 - (x - x_1)^3 M_2\right]$$

$$+(x_2 - x)\left[y_1 - \frac{M_1}{6}\right] + (x - x_1)\left(y_2 - \frac{M_2}{6}\right)$$

$$= \frac{1}{6}(2 - x)^3 (33) + (2 - x)\left(3 - \frac{33}{6}\right) + (x - 1)(29 - 0)$$

$$f(x) = S(x) = \frac{11}{2}(2 - x)^3 - \frac{5}{2}(2 - x) + 29(x - 1) \text{ in } (1 \le x \le 2) \qquad \ldots(3)$$

Put x = 0.5 in (2), 0.5 in (0, 1):

$$f(0.5) = \frac{11}{2}(0.5)^3 - \frac{3}{2}(0.5) - 1$$

$$f(0.5) = -1.0625.$$

$$f(1.5) = \frac{11}{2}(0.5)^3 - \frac{5}{2}(0.5) + 29(0.5)$$

$$f(1.5) = 13.94$$

5. Let us find the cubic splines and evaluate y(1 .5) and y (3).

The following values of x and y are given,

x:	1	2	3	4
y:	1	2	5	11

Solution:

Since the values of x are equally spaced with:

$$h = 1,\ x_0 = 1;\ x_1 = 2,\ x_2 = 3,\ x_3 = 4;$$

$$y_0 = 1, y_1 = 2; y_2 = 5, y_3 = 11$$

$$M_{i-1} + 4M_i + M_{i+1} = 6(y_{i-1} - 2y_i + y_{i+1}) \quad i = 1, 2 \qquad ...(1)$$

And $M_0 = 0; M_3 = 0$

Put i = 1 in (1),

$$M_0 + 4M_1 + M_2 = 6(y_0 - 2y_1 + y_2)$$

$$0 + 4M_1 + M_2 = 6(1 - 4 + 5)$$

i. e., $4M_1 + M_2 = 12 \qquad ...(2)$

Put i = 2 in (1):

$$M_1 + 4M_2 + M_3 = 6(y_1 - 2y_2 + y_3)$$

$$M_1 + 4M_2 + 0 = 6(2 - 10 + 11)$$

$$M_1 + 4M_2 = 18 \qquad ...(3)$$

Solving (2) and (3) we get:

$$M_1 = 2 \text{ and } M_2 = 4 \text{ and } M_0 = M_3 = 0$$

The cubic spline in $x_{i-1} \le x \le x_i$ is given by:

$$y = \frac{1}{6}\left[(x_i - x)^3 M_{i-1} + (x - x_{i-1})^3 M_i\right]$$

$$+(x_i - x)\left(y_{i-1} - \frac{1}{6}M_{i-1}\right) + (x - x_{i-1})\left(y_i - \frac{1}{6}M_i\right) \qquad ...(4)$$

- Put i = 1 in (4): Cubic spline in (1,2) is given by,

$$y = \frac{1}{6}\left[(x_1 - x)^3 M_0 + (x - x_0)^3 M_1\right] + (x_1 - x)\left(y_0 - \frac{M_0}{6}\right) + (x - x_0)\left(y_1 - \frac{M_1}{6}\right)$$

$$y = \frac{1}{6}\left[0 + 2(x-1)^3\right] + 2(2 - x)(1 - 0) + (x - 1)\left(2 - \frac{2}{6}\right)$$

$$y = \frac{1}{3}(x-1)^3 + (2 - x) + \frac{5}{3}(x - 1) \text{ in } (1,2).$$

- Put i = 2 in (4), we get cubic spline in (2,3):

$$y=\frac{1}{6}\left[(x_2-x)^3 M_1+(x-x_1)^3 M_2\right]+(x_2-x)\left(y_1-\frac{M_1}{6}\right)+(x-x_1)\left(y_2-\frac{M_2}{6}\right)$$

$$y=\frac{1}{6}\left[2(3-x)^3+4(x-2)^3\right]+(3-x)\left(2-\frac{2}{6}\right)+(x-2)\left(5-\frac{4}{6}\right)$$

$$y=\frac{1}{3}(3-x)^3+\frac{2}{3}(x-2)^3+\frac{5}{3}(3-x)+\frac{13}{3}(x-2)\text{in}(2,3).$$

Put i = 3 in (4):

$$y=\frac{1}{6}\left[(x_3-x)^3 M_2+(x-x_2)^3 M_3\right]+(x_3-x)\left(y_2-\frac{M_2}{6}\right)+(x-x_2)\left(y_3-\frac{M_3}{6}\right)$$

$$y=\frac{1}{6}\left[4(4-x)^3+0\right]+(4-x)+\left(5-\frac{4}{6}\right)+(x-3)(11)$$

$$=\frac{2}{3}(4-x)^3+\frac{13}{3}(4-x)+11(x-3)\text{in}(3,4).$$

∴ The cubic spline is:

$$y(x)=\frac{1}{3}(x-1)^3+(2-x)+\frac{5}{3}(x-1) \text{ in } (1,2) \quad ...(5)$$

$$y(x)=\frac{1}{3}(3-x)^3+\frac{2}{3}(x-2)^3+\frac{5}{3}(3-x)+\frac{13}{3}(x-2) \text{ in } (2,3) \quad (6) \quad ...(6)$$

$$y(x)=\frac{2}{3}(4-x)^3+\frac{13}{3}(4-x)+11(x-3) \text{ in } (3,4) \quad ...(7)$$

Different (7) with respect to *x*:

$$y(x)=2(4-x)^2(-1)-\frac{13}{3}+11 \text{ in } (3,4) \quad ...(8)$$

Put x = 1.5 in (5):

$$y(1.5)=\frac{1}{3}(1.5-1)^3+(2-1.5)+\frac{5}{3}(1.5-1)$$

$$y(1.5)=1.375$$

$$y(3)=2(4-3)^2(-1)-\frac{13}{3}+11$$

$$=-2-\frac{11}{3}+11$$

$$y(3)=5.333.$$

1.5 Fitting a Straight Line: Second Degree Curve, Exponential Curve, Power Curve by Method of Least Squares

Least Square Fit of a Straight Line

Suppose that we are given a data set $(x_1,y_1),(x_2,y_2),(x_3,y_3),\ldots,(x_n,y_n)$ of n observations from an experiment. Say that we are interested in fitting a straight line:

$$y=ax+b$$

to the given data. Determine the 'n' residuals e_i by:

$$e_i=y_i-(ax_i+b),\quad i=1,2\ldots n \qquad \ldots(2)$$

Now consider the sum of the squares of e_is i.e:

$$E=\sum_{i=1}^{n}e_i^2$$

$$E=\sum_{i=1}^{n}\left[y_i-(ax_i+b)^2\right] \qquad \ldots(3)$$

Note that E is a function of parameters a and b. We required to find a,b such that E is minimum. The necessary condition for E to be minimum is given by:

$$\frac{\partial E}{\partial a}=\frac{\partial E}{\partial b}=0 \qquad \ldots(4)$$

The condition $\frac{\partial E}{\partial a}=0$ yields:

$$\frac{\partial E}{\partial a}=\sum_{i=1}^{n}2x_i\left[y_i-(ax_1+b)\right]=0$$

$$\text{i.e } a\sum_{i=1}^{n} x_i^2 + b\sum_{i=1}^{n} x_i y_i = \sum_{i=1}^{n} x_i y_i \qquad ...(5)$$

Similarly the condition $\frac{\partial E}{\partial a} = 0$ yields:

$$a\sum_{i=1}^{n} x_i + nb = \sum_{i=1}^{n} y_i \qquad ...(6)$$

Equations (5) and (6) are known as normal equations, which are to be solved to get desired values for a and b.

The expression for E i.e (3) will be re-written in a convenient way as follows:

$$E = \left(\sum_{i=1}^{n} y_i^2 - a\sum_{i=1}^{n} x_i y_i - b\sum_{i=1}^{n} y_i\right) \qquad ...(7)$$

Problem

1. Using the method of least squares, let us find an equation of the form $y = ax + b$ that fits the following data:

x	0	1	2	3	4
y	1	5	10	22	38

Solution:

Consider the normal equations of least square fit of a straight line i.e:

$$a\sum_{i=1}^{n} x_i^2 + b\sum_{i=1}^{n} x_i = \sum_{i=1}^{n} x_i y_i \qquad ...(1)$$

$$a\sum_{i=1}^{n} x_i + nb = \sum_{i=1}^{n} y_i \qquad ...(2)$$

Here n=5

From the given data, we have:

x	y	xy	x^2
0	1	0	0
1	5	5	1
2	10	20	4
3	22	66	9
4	38	152	16

$$\sum_{i}^{x=10}\sum_{i} y_i = 76 \sum_{i} x_i y_i = 243 \sum_{i} x_i^2 = 30$$

Therefore, the normal equations are given by:

$$30a + 10b = 243 \quad \text{...(3)}$$

$$10a + 5b = 76 \quad \text{...(4)}$$

On solving (3) and (4) we get:

$$a = 9.1,\ b = -3 \quad \text{...(5)}$$

Hence, the needed fit for the given data is:

$$y = 9.1x - 3$$

Second Degree Curve

The least-squares parabola uses a second degree curve:

$$y = a + bx + cx^2$$

To approximate the given set of data, $(x_1, y_1), (x_2, y_2), \ldots, (x_n, y_n)$, where $n \geq 3$. The best fitting curve $f(x)$ has the least square error, i.e.:

$$\Pi = \sum_{i=1}^{n}[y_i - f(x_i)]^2 = \sum_{i=1}^{n}\left[y_i - \left(a + bx_i + cx_i^2\right)\right]^2 = \min$$

Please note that a, b and c are unknown coefficients while all x_i and y_i are given. To obtain the least square error, the unknown coefficients a, b, and c should yield zero first derivatives:

$$\begin{cases} \dfrac{\partial \Pi}{\partial a} = 2\sum_{i=1}^{n}\left[y_i - \left(a + bx_i + cx_i^2\right)\right] = 0 \\ \dfrac{\partial \Pi}{\partial b} = 2\sum_{i=1}^{n}x_i\left[y_i - \left(a + bx_i + cx_i^2\right)\right] = 0 \\ \dfrac{\partial \Pi}{\partial c} = 2\sum_{i=1}^{n}x_i^2\left[y_i - \left(a + bx_i + cx_i^2\right)\right] = 0 \end{cases}$$

Expanding the above equations, we have:

$$\begin{cases} \sum_{i=1}^{n} y_i = a\sum_{i=1}^{n} 1 + b\sum_{i=1}^{n} x_i + c\sum_{i=1}^{n} x_i^2 \\ \sum_{i=1}^{n} x_i y_i = a\sum_{i=1}^{n} x_i + b\sum_{i=1}^{n} x_i^2 + c\sum_{i=1}^{n} x_i^3 \\ \sum_{i=1}^{n} x_i^2 y_i = a\sum_{i=1}^{n} x_i^2 + b\sum_{i=1}^{n} x_i^3 + c\sum_{i=1}^{n} x_i^4 \end{cases}$$

The unknown coefficients a, b, and e will hence be obtained by solving the above linear equations.

Exponential Curve Method

Suppose that we are interested in fitting a curve of the form:

$$y = a_0\left(1 - e^{-a_1 x}\right) \qquad ...(1)$$

To a given data set of n observations from an experiment. Unlike the transcendental curve or exponential curve one cannot linearize (1) by taking logarithms on both the sides. But, like in linear regression analysis, nonlinear regression is based on determining the values of the parameters that minimize the sum of the squares of residuals. In the nonlinear case this is achieved in a iterative fashion.

The Gauss-Newton method is used for minimizing the sum of the squares of the residuals between data and nonlinear equations. Here the Taylor series expansion is used to express the original nonlinear equation in an approximate, linear form. Then the principle of least square is used to obtain new estimates of the parameters that move in the direction of minimizing the residual.

Now to illustrate the methodology let us represent (1) as:

$$y = f(x, a_0, a_1) \qquad ...(2)$$

Let us suppose that we are given a data set $(x_1, y_1), (x_2, y_2)$, (x_n, y_n) of n observations from an experiment. At each of (x_i, y_i) while y_i is the measured value, $f(x_i, a_0, a_1)$ can be the estimated value. For simplicity we denote $f(x_i, a_0, a_1)$ by $f(x_i)$. Now going by the Gauss-Newton method, let us linearize the nonlinear mode at $(j+1)^{th}$ iteration level using Taylor series as follows:

$$f(x_i)_{j+1} \simeq f(x_i)_j + \frac{\partial f(x_i)_j}{\partial a_0}\Delta a_0 + \frac{\partial f(x_i)_j}{\partial a_1}\Delta a_1 \qquad ...(3)$$

Where the subscripts $j,(j+1)$ denote iteration levels:

$$\Delta a_o = a_{o,j+1} - a_{o,j} \text{ and } \Delta a_1 = a_{1,j+1} - a_{1,j} \qquad ...(4)$$

Now we may note that (3) represents the linearized version of the model w.r.t iteration level. i.e, at $j,(j+1)^{th}$ iteration we already know the jth level values of the parameters i.e $a_{o,j}$ & $a_{1,j}$. Now we have the residuals e, given by:

$$e_i = y_j - f(x_i)_j \simeq \frac{\partial f(x_i)_j}{\partial a_o} \Delta a_o + \frac{\partial f(x_i)_j}{\partial a_1} \Delta a_1 \qquad ...(5)$$

$i = 1,2..n$

Which in the matrix form may be written as:

$$\{D\} = [Z_j]\{\Delta A\} \qquad ...(6)$$

Where,

$$[Z_j] = \begin{bmatrix} \frac{\partial f(x_1)_j}{\partial_{ao}} & \frac{\partial f(x_1)_j}{\partial_{a1}} \\ \frac{\partial f(x_2)_j}{\partial_{ao}} & \frac{\partial f(x_2)_j}{\partial_{a1}} \\ \cdots & \cdots \\ \frac{\partial f(x_n)_j}{\partial_{ao}} & \frac{\partial f(x_n)_j}{\partial_{a1}} \end{bmatrix} \qquad ...(7.1)$$

i.e., $[Z_j]$ is the matrix of partial derivatives of the function evaluated with the jth level iteration values.

$$\{D\} = \begin{bmatrix} y_1 & - & f(x_1)_j \\ y_2 & - & f(x_2)_j \\ \cdots & & \\ y_n & - & f(x_n)_j \end{bmatrix} \{\Delta A\} = \begin{bmatrix} \Delta a_o \\ \Delta a_1 \end{bmatrix} \qquad ...(7.2)$$

i.e., $\{D\}$ contains the differences between the measurements and the function values and $\{\Delta A\}$ contains the changes in the parameter values.

Applying the principle of least squares on (6) we arrive at:

$$\left[\left[Z_j\right]^T\left[Z_j\right]\right]\{\Delta A\}=\left\{\left[Z_j\right]^T\{D\}\right\} \quad ...(8)$$

Now on solving (8) we obtain $\{\Delta A\}$, which can be used to compute improved values of the parameters a_o, a_1 i.e.:

$$a_{o,j+1}=a_{o,j}+\Delta a_o \quad ...(9.1)$$

$$a_{1,j+1}=a_{1,j}+\Delta a_1 \quad ...(9.2)$$

We repeat the above procedure until the solution converges i.e. until:

$$\left|\varepsilon_a\right|_k=\left|\frac{a_{k,j+1}-a_{k,j}}{a_{k,j+1}}\right|<\varepsilon^k_{tolerance} \qquad k=0,1 \quad ...(10)$$

Where $\left|\varepsilon_a\right|_k$ denotes the relative error in a^{th}_k parameter and $\varepsilon^k_{tolerance}$ is some prescribed tolerance level for the relative error in a^{th}_k parameter.

Power Curve Fit

Suppose we have data that, when plotted, appear to have a power-law character. If we choose a power function to represent the data, we write:

$$y=ax^b$$

As in the previous section, a and b are the λ m parameters we seek that would best fit the function to the data. Thus, we write for the total squared error:

$$E^2=\sum_{i-1}^N \varepsilon_i^2=\sum_{i-1}^N\left[y_i-f(x_i)\right]^2=\sum_{i-1}^N\left[y_i-ax_i^b\right]^2$$

The two parameters are found by:

$\frac{\partial E^2}{\partial a}=0$ and $\frac{\partial E^2}{\partial b}=0$ which are solved simultaneously to obtain:

$$b=\frac{\sum \text{In}\, x_i\, \text{In}(y_i)-\frac{1}{N}\left(\sum \text{In}\, x_i\right)\left(\sum \text{In}(y_i)\right)}{\left(\sum(\text{In}\, x_i)\right)^2-\frac{1}{N}\left(\sum \text{Inx}_i\right)^2}$$

$$a=\exp\left[\frac{1}{n}\text{In}\, y_i-b\frac{\sum \text{Inx}_i}{N}\right]$$

It can be shown that this yields a Coefficient of Determination of:

$$r^2 = \frac{\left[\sum \text{In}\,x_i\,\text{In}\,y_i - \frac{1}{N}\sum \text{In}\,x_i \sum \text{In}\,y_i\right]^2}{\left[\sum(\text{In}\,x_i)^2 - \frac{\left(\sum \text{In}\,x_i\right)^2}{N}\right]\left[\sum(\text{In}\,y_i)^2 - \frac{\left(\sum \text{In}\,y_i\right)^2}{N}\right]}$$

2

Numerical Techniques

2.1 Introduction

The problem of solving the equation f(x) =0 is of great importance in Science and Engineering. In this section, we deal with the various methods which give a solution for the equation f(x) =0.

Solution of Algebraic and Transcendental Equations

The equation of the form f(x) =0 are called algebraic equations, if f(x) is purely a polynomial in x. For example: $2x^3 - 3x^2 - x + 5 = 0$ & $x^4 - x - 10 = 0$ are algebraic equations.

If f(x) also contains trigonometric, logarithmic, exponential function etc. then the equation f(x) =0 is known as transcendental equation.

Methods for Solving the Equation f(x) =0

The following result helps us to locate the interval in which the roots of are:

- Method of false position.
- Iteration method.
- Newton-Raphson method.

2.1.1 Graphical Interpretation of Solution of Equations

In order to solve graphically an equation:

$$f(x) = 0,$$

It is necessary to find the points at which a curve:

$$y = f(x)$$

Intersects or touches the x-axis. The x-coordinates of a graph, that are common with the x-axis, are called the x-intercepts.

In order to find the x-intercepts of the function y = f (x), set y = 0 and solve the equation f (x) = 0 for x. Solutions of an equation f (x) = 0 are interpreted graphically as the x-intercepts of the curve y = f (x).

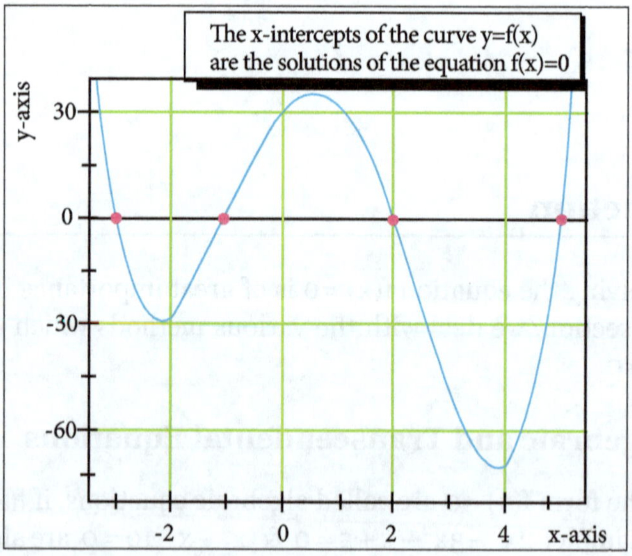

The simplest algebraic equation is a linear equation:

$$kx+b=0$$

Where x is the variable; k and b are constants (k ≠ 0).

A graph of a linear function:

$$y=kx+b=0\ (k\neq 0)$$

Is a straight line with the only one x-intercept, x = – b/a.

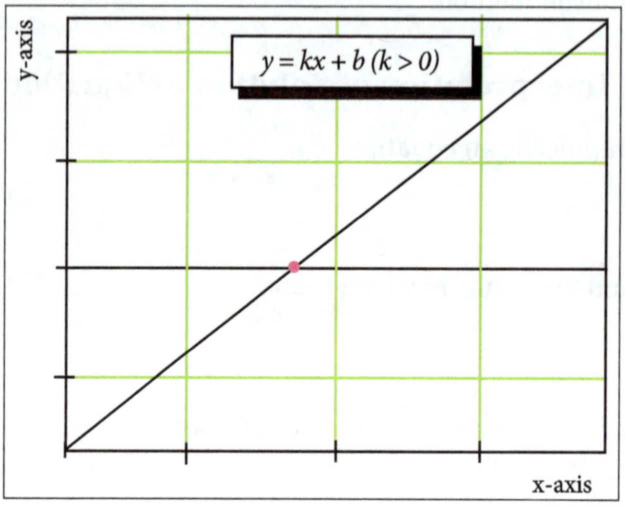

So a linear equation has a single solution:

$$x = -\frac{b}{k}$$

In mathematical literature, the equation k x + b = 0 is known as an equation of a line in slope-intercept form, since b is the y-intercept, and k is the slope of the line.

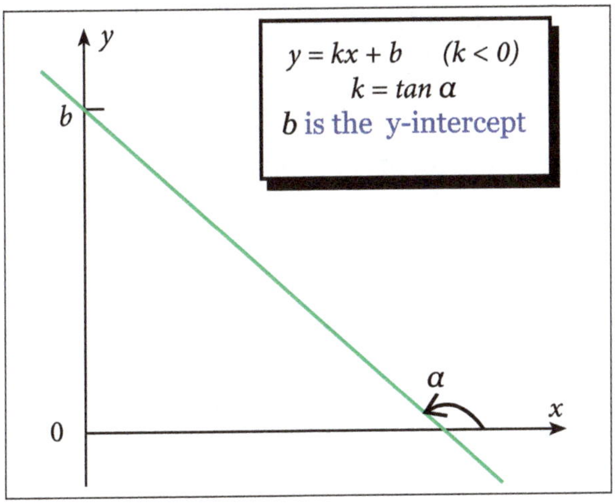

A linear inequality in one variable can be put into one of the following forms:

$$a\,x + b > 0 (*)$$

or,

$$a\,x + b \geq 0, (**)$$

Where a and b are constants (a ≠ 0); x is the variable.

Inequality (*) is known as a strict inequality, while inequality (**) is an unstrict inequality. The only difference between them is whether the endpoint of the interval is included in the solution set or not.

The solution set for inequality (*):

$$x > -\frac{b}{a}$$

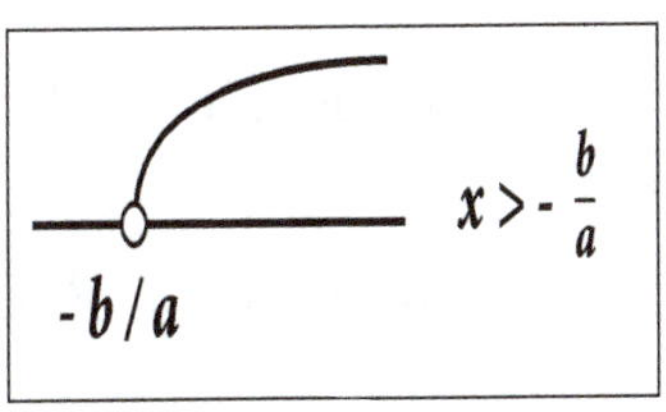

The solution set for inequality (**):

$$x \geq -\frac{b}{a}$$

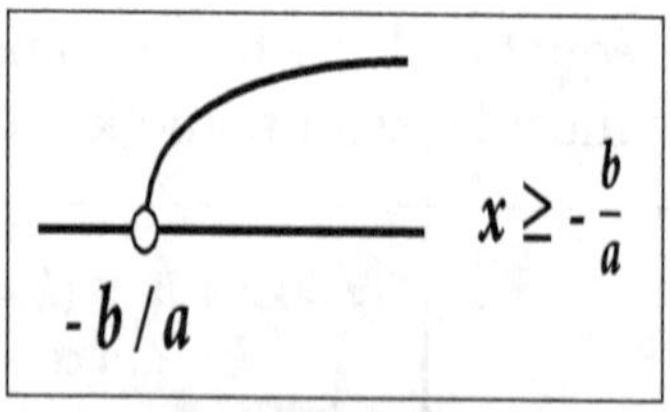

In order to solve an equation involving the absolute value $|a\,x + b|$, it is necessary to consider two cases.

Case 1:

If the expression $a\,x + b$ represents a positive quantity, then the absolute value symbol can be simply dropped:

$$a\,x + b \geq 0 \Rightarrow |a\,x + b| = a\,x + b.$$

Case 2:

If the expression $a\,x + b$ represents a negative quantity, then

$$a\,x + b < 0 \Rightarrow |a\,x + b| = -(a\,x + b)$$

Thus, to solve a linear equation involving the absolute value $|a\,x + b|$ means to solve two ordinary linear equations.

2.1.2 The Bisection Method

The Bisection Method is a numerical method for estimating the roots of a polynomial f(x). It is one of the simplest and most reliable but it is not the fastest method. Assume that f(x) is continuous.

Algorithm for the Bisection Method: Given a continuous function f(x):

- Find points a and b such that a < b and f(a) * f(b) < 0.
- Take the interval [a, b] and find its midpoint x_1.
- If $f(x_1) = 0$ then x_1 is an exact root, else if $f(x_1) * f(b) < 0$ then let $a = x_1$, else if $f(a) * f(x_1) < 0$ then let $b = x_1$.
- Repeat steps 2 & 3 until $f(x_i) = 0$ or $|f(x_i)| <= DOA$, where DOA stands for degree of accuracy.

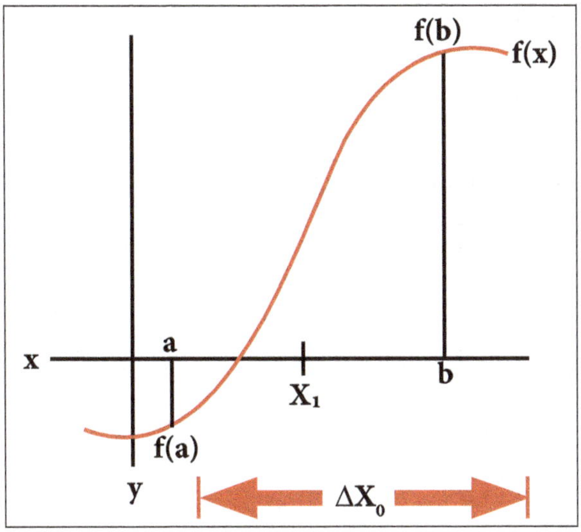

For the ith iteration, where i = 1, 2, . . . , n, the interval width is:

$$\Delta x_i = 0.5\ \Delta x_{i-1} = (0.5)i(b-a)$$

And the new midpoint is,

$$x_i = a_{i-1} + \Delta x_i$$

Problems

1. We shall show that $f(x) = x^3 + 4x^2 - 10 = 0$ has a root in [1, 2] and by using the Bisection method to determine an approximation to the root that is accurate to at least within 10^{-4}.

Solution:

For this example, the iteration will be terminated when a bound for the relative error is less than 10^{-4}, implemented in the form:

$$\frac{|p_n - p_{n-1}|}{|p_n|} < 10^{-4}$$

Because f(1) = −5 and f(2) = 14 the Intermediate Value Theorem ensures that this continuous function has a root in [1, 2].

For the first iteration of the Bisection method we use the fact that at the midpoint of [1, 2] we have f(1.5) = 2.375 > 0.

This indicates that we should select the interval [1, 1.5] for our second iteration.

Then we find that f(1.25) = −1.796875 so our new interval becomes [1.25, 1.5], whose midpoint is 1.375. Continuing in this manner gives the values shown in the following table:

Iter	a_n	b_n	p_n	$f(a_n)$	$f(p_n)$	RelErr
1	1.000000	2.000000	1.500000	−5.000	2.375	0.33333
2	1.000000	1.500000	1.250000	−5.000	−1.797	0.20000
3	1.250000	1.500000	1.375000	−1.797	0.162	0.09091
4	1.250000	1.375000	1.312500	−1.797	−0.848	0.04762
5	1.312500	1.375000	1.343750	−0.848	−0.351	0.02326
6	1.343750	1.375000	1.359375	−0.351	−0.096	0.01149
7	1.359375	1.375000	1.367188	−0.096	0.032	0.00571
8	1.359375	1.367188	1.363281	−0.096	−0.032	0.00287
9	1.363281	1.367188	1.365234	−0.032	0.000	0.00143
10	1.363281	1.365234	1.364258	−0.032	−0.016	0.00072
11	1.364258	1.365234	1.364746	−0.016	−0.008	0.00036
12	1.364746	1.365234	1.364990	−0.008	−0.004	0.00018
13	1.364990	1.365234	1.365112	−0.004	−0.002	0.00009

After 13 iterations, $p_{13} = 1.365112305$ approximates the root p with an error:

$$|p - p_{13}| < |b_{14} - a_{14}| = |1.3652344 - 1.3651123| = 0.0001221$$

Since,

$$|a_{14}| < |p|,$$

We have,

$$\frac{|p - p_{13}|}{|p|} < \frac{|b_{14} - a_{14}|}{|a_{14}|} \leq 9.0 \times 10^{-5}.$$

So the approximation is correct to at least within 10^{-4}.

The correct value of p to nine decimal places is $p = 1.365230013$

2. Let us calculate the root of $x^4 - x - 10 = 0$ using bisection method.

The graph of this equation is given in the figure.

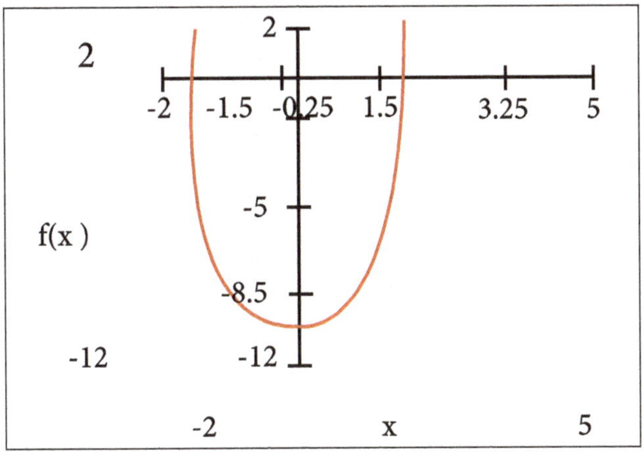

Let a = 1.5 and b = 2

Iteration No.	a	b	c	f(a) * f(c)
1	1.5	2	1.75	15.264 (+ve)
2	1.75	2	1.875	−1.149 (−ve)
3	1.75	1.875	1.812	2.419 (+ve)
4	1.812	1.875	1.844	0.303 (−ve)
5	1.844	1.875	1.86	−0.027 (−ve)

So one of the roots of $x^4 - x - 10 = 0$ is approximately 1.86.

3. Let us calculate the root of $x - \sin(x) - (1/2) = 0$.

The graph of this equation is given in the figure.

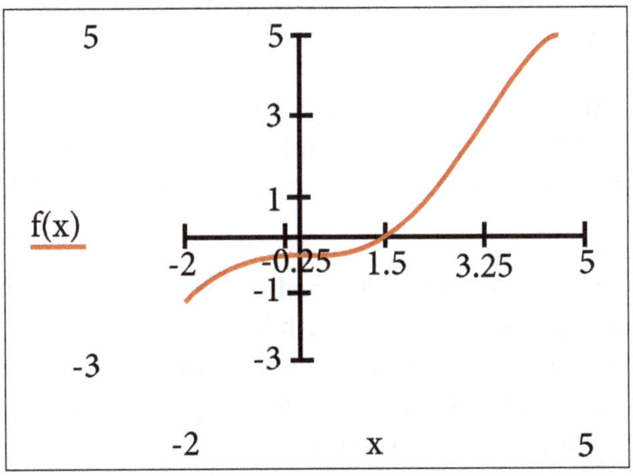

Let a = 1 and b = 2

Iteration No.	a	b	c	f(a) * f(c)
1	1	2	1.5	$-8.554*10^{-4}$ (-ve)
2	1	1.5	1.25	0.068 (+ve)
3	1.25	1.5	1.375	0.021 (+ve)
4	1.375	1.5	1.437	$5.679*10^{-3}$ (+ve)
5	1.437	1.5	1.469	$1.42*10^{-3}$ (+ve)
6	1.469	1.5	1.485	3.042*10–4 (+ve)
7	1.485	1.5	1.493	5.023*10–5 (+ve)
8	1.493	1.5	1.497	$2.947*10^{-6}$ (+ve)

So one of the roots of x-sin(x)-(1/2) = 0 is approximately 1.497.

Advantages and disadvantages of the bisection method:

- The method is guaranteed to converge.
- The error bound decreases by half with each iteration.
- The bisection method converges very slowly.
- The bisection method cannot detect multiple roots.

2.2 The Method of False Position

Method of False Position or Regula-Falsi Method or Linear Interpolation Method

In bisection method, the interval is always divided into half. If a function changes sign over an interval, the function value at the mid-point is evaluated. In bisection method, the interval from a to b into equal intervals, no account is taken of the magnitude of f(a) and f(b).

An alternative method that exploits this graphical insights is to join f(a) and f(b) by a straight line. The intersection of this line with the X-axis represents an improved estimate of the root. The replacement of the curve by a straight line gives a “false position” of the root is the origin of the name, method of false position or in Latin Regula-falsi. It is also called the linear interpolation method.

Problems

1. Let us find a real root of $x^3 - 9x + 1 = 0$ that lies between 2.5 and 3 by the method of false position and correct to three decimal places.

Solution:

Let,

$$f(x) = x^3 - 9x + 1 = 0$$

$$f(2.5) = -5.875 \text{ and } f(3) = 1$$

The root lies between 2.5 and 3.

Iteration(r)	A	b	x_r	$f(x_r)$
1	2.5	3	2.9273	−0.02664
2	2.927	3	2.9423	−0.0088
3	2.942	3	2.9425	−0.0054
4	2.942	3	2.9425	0.003

The approximations are given by:

$$x = \frac{af(b) - bf(a)}{f(b) - f(a)}$$

2. Obtain the real root of $x^x = 100$, correct to four decimal places using the method of false position.

Solution:

Given:

$$x^x = 100$$
$$x \log_{10} x = 2$$
$$f(x) = x \log_{10} x - 2$$
$$f(3{,}5) = -0.0958 < 0 \text{ and } f(3.6) = 0.00027 >$$

Taking logarithmic on both sides:

The roots lie between 3.5 and 3.6.

The approximation is given by:

$$x = \frac{af(b) - bf(a)}{f(b) - f(a)}$$

Iteration (r)	a	b	x_r	$f(x_r)$
1	3.5	3.6	3.5973	0.000015
2	3.5	3.5973	3.5973	0

The required root is 3.5973.

2.2.1 The Iteration Method

For solving the equation f(x) = 0 by iteration method, we start with an approximation value of the root. The equation f(x) = 0 is expressed as x = φ(x). The equation x = φ(x) is called fixed point equation.

The iteration formula is given by, $x_{n+1} = \phi(x_n)$, $n = 0,1,2,...$ and it is called as fixed point iteration formula.

Let f(x) = 0 be the given equation whose exact root is α.

The equation f(x) = 0 be rewritten as, x = φ(x).

Let I be the interval containing the root x = α. If $|\phi'(x) < 1|$ for all x in I, then the sequence of approximation $x_1, x_2, x_3 .... x_n$ will converges to α. If the initial starting value x_0 is chosen in I.

Problems

1. Let us solve $e^x - 3x = 0$ by the method of fixed point iteration.

Given:

$$e^x - 3x = 0$$

$$f(x) = e^x - 3x$$

$$f(0) = 1(+ve)$$

$$f(1) = -0.2817\ (-ve)$$

A root lies between 0 and 1, which is nearest to 1.

The given equation can be written as:

$$x = \frac{e^x}{3} = \phi(x)$$

$$\phi(x) = \frac{e^x}{3} \;\&\; |\phi(x)| < 1 \text{ in } (0,1)$$

Choose $x_0 = 0.7$

$$x_{n+1} = \phi(x_0)$$

$$x_1 = \phi(x_0) = \frac{e^x o}{3} = \frac{e^{0.7}}{3} = 0.67125$$

$$x_2 = \frac{e^{x_1}}{3} = 0.65223$$

$$x_3 = \frac{e^{x_2}}{3} = 0.63994$$

$$x_4 = 0.63212$$

$$x_5 = 0.6272$$

$$x_6 = 0.62412$$

$$x_7 = 0.6222$$

$$x_8 = 0.62101$$

$$x_9 = 0.62027$$

$$x_{10} = 0.61981$$

$$x_{11} = 0.61952$$

$$x_{12} = 0.61935$$

$$x_{13} = 0.61924$$

$$x_{14} = 0.6192$$

$$x_{15} = 0.6192$$

Here,

$$x_{14} = x_{15} = 0.6192$$

$$\therefore x = 0.6192$$

2. Let us find a positive root of the equation cos x - 3x + 1 = 0 by using iteration method.

Solution:

Let,

$$f(x) = \cos x - 3x + 1 = 0$$
$$f(0) = 1 - 0 + 1 = 2 \ (+ve)$$
$$f(1) = -1.4597 \ (-ve)$$

Therefore, a root lies between 0 and 1.

Here,

$$|f(1)| < |f(0)|$$

Hence root is nearer to 1.

The given equation may be written as:

$$x = \left(\frac{1+\cos x}{3}\right) = \phi(x)$$
$$\phi(x) = -\frac{1}{3}\sin x \text{ and } |\phi(x)| < 1 \text{ in } (0,1)$$

Hence the iteration method is applicable.

Iteration formula:

$$xn + 1 = x_m \quad \text{...(1)}$$

Let,

$$x_0 = 0.6.$$

$$x_1 = \phi(x_0) = \frac{1}{3}(1 + \cos x_0) = \frac{1}{3}(1 + \cos(0.6))$$

$$= 0.60845$$

$$x_2 = \phi(x_1) = \frac{1}{3}(1 + \cos x_1)$$
$$= 0.60684$$

$$x_3 = 0.60715$$
$$x_4 = 0.60709$$
$$x_5 = 0.60710$$
$$x_6 = 0.60710$$

Here,

$$x_5 = x_6 = 0.6071.$$

Hence the root is x = 0.6071.

2.2.2 Newton-Raphson Method

This method starts with an initial approximation to the root of an equation, a better and closer approximation to the root can be found by using an iterative process.

Derivation of Newton-Raphson Formula

Let α be the root of f(x) = 0 and x_0 be an approximation to α.

If,

$$h = \alpha - X_0.$$

Then by the Taylor's series is:

$$x_{i+1} = x_i - \frac{f(x_i)}{f'(x_i)}, \ i = 0,1,2,3,.....$$

Condition for Convergence of Newton-Raphson Method for f(x) = 0

Let f(x) = 0, if the root lies between (a, b) then condition for convergence of Newton Raphson method is $|f(x)\ddot{f}(x)| < |\dot{f}(x)|^2$ in the internal, (a, b).

Problems

1. Using Newton's method, let us find the root between 0 and 1 of $x^3 - 6x + 4 = 0$ correct five decimal places.

Solution:

Given:

$$x^3 - 6x + 4 = 0$$

$$f(x) = x^3 - 6x + 4$$

$$f'(x) = 3x^2 - 6$$

By Newton-Raphson formula, we have approximation:

$$x_{i+1} = x_i - \frac{f(x_i)}{f'(x_i)}, \; n = 0,1,2,3,\ldots\ldots$$

The initial approximation is $x_0 = 0.5$.

First approximation:

$$x_1 = x_0 - \frac{f(x_0)}{f'(x_0)},$$
$$= 0.5 - \frac{1.125}{-5.25},$$
$$= 0.71429$$

Second approximation:

$$x_2 = x_1 - \frac{f(x_1)}{f'(x_1)},$$
$$= 0.71429 - \frac{0.0787}{-4.4694},$$
$$= 0.7319$$

Third approximation:

$$x_1 = x_2 - \frac{f(x_2)}{f'(x_2)},$$
$$= 0.73205 - \frac{0.0006}{-4.3923},$$
$$= 0.73205$$

Fourth approximation:

$$x_4 = x_3 - \frac{f(x_3)}{f'(x_3)},$$
$$= 0.73205 - \frac{0.0006}{-4.3923},$$
$$= 0.73205$$

The root is 0.73205, correct to five decimal places.

2. Let us evaluate $\sqrt{15}$ using Newton-Raphson's formula.

Let,

$$x=\sqrt{15}$$

$$\therefore \quad x^2=15$$

Solution:

$$x^2-15=0$$

i.e., $f(x)=x2-15;\ f(x)=2x.$

$$x_{n+1}=x_n-\frac{f(x_n)}{f(x_n)}$$

$$=x_n-\frac{(x^2-15)}{2x_n}=\frac{x_n^2+15}{2x_n} \qquad ...(1)$$

Take xo = 4. [∵ f(3) & f(4) are of opposite sign root lies between 3 and 4].

$$x_1=\frac{x_0^2+15}{2x_0}=3.875$$

$$x_2=\frac{x_1^2+15}{2x_1}=3.873$$

$$x_3=3.873$$

$$\therefore \quad x=\sqrt{15}=3.873$$

3. Using Newton's method, let us find the root between 0 and 1 of $x^3 = 6x - 4$.

Solution:

Given:

$$x^3-6x+4=0$$

Here,

$$f(x)-x^3-6x+4$$

$$f(x)=3x^2-6$$

$$x_{n+1} = x_n - \frac{f(x_n)}{f(x_n)}$$

$$= x_n - \frac{(x_n^3 - 6x_n + 4)}{3x_n^2 - 6}$$

$$= \frac{2x_n^3 - 4}{3x_n^2 - 6}$$

Newton's iterative formula is:

Let,

$$x_0 = 1$$

$$x_1 = \frac{2x_0^3 - 4}{3x_0^2 - 6} = 0.666$$

$$x_2 = \frac{2x_1^3 - 4}{3x_1^2 - 6} = 0.7301$$

$$x_4 = 0.732$$

$$x_3 = x_4$$

$$x = 0.732$$

4. Using Newton-Raphson method, let us find a positive root of the equation 3x - cos x - 1 = 0.

Solution:

Let,

$$f(x) = 3x - \cos x - 1;\ f'(x) = 3 + \sin x$$

$$f(0) = -2\ (-ve)$$

$$f(1) = 1.4597\ (+ve)$$

∴ The root lies between 0 and 1.

Here $|f(1)| < |f(0)|$.

Hence the root is nearer to 1.

$$x_{n+1} = x_n - \frac{f(x_n)}{f(x_n)}$$

$$x_{n+1} = x_n - \frac{(3x_n - \cos x_n - 1)}{(3 + \sin x_n)}$$

$$x_{n+1} = \frac{x_n \sin x_n + \cos x_n + 1}{3 + \sin x_n} \qquad ...(1)$$

Newton iterative formula is:

Let,

$$x_o = 0.6$$

Put,

n = o in (I) we get:

$$x_1 = \frac{0.6\sin(0.6) + \cos(0.6) + 1}{3 + \sin(0.6)} = 0.6071$$

Put,

n = 1 in (I),

$$x_2 = \frac{(0.6071)\sin(0.671) + \cos(0.6071) + 1}{3 + \sin(0.6071)} = 0.6071$$

Here,

$$x_1 = x_2 = 0.6071.$$

Hence the root is $x = 0.6071$.

2.3 Solving System of Non-Homogeneous Equations

These methods are used to solve a special of linear equations in which each equation must possess one large coefficient and the large coefficient must be attached to a different unknown in that equation.

Further in each equation, the absolute value of the large coefficient of the unknown is greater than the sum of the absolute values of the other coefficients of the other unknowns.

Such type of simultaneous linear equations can be solved by the following iterative methods:

- Gauss-Jacobi method.
- Gauss seidel method.

$$a_1x + b_1y + c_1z = d_1$$
$$a_2x + b_2y + c_2z = d_2$$
$$a_3x + b_3y + c_3z = d_3$$

i.e., the co-efficient matrix $A = \begin{bmatrix} \langle a_1 \rangle & b_1 & c_1 \\ a_2 & \langle b_2 \rangle & c_2 \\ a_3 & b_3 & \langle c_3 \rangle \end{bmatrix}$ is diagonally dominant.

$$x = \frac{1}{a_1}[d_1 - b_1y - c_1z]$$

$$y = \frac{1}{b_2}[d_2 - a_2x - c_2z]$$

$$z = \frac{1}{c_3}[d_3 - a_3x - b_3y]$$

Gauss-Jacobi Method

If the rth iterates are $x^{(r)}, y^{(r)}, z^{(r)}$ then the iteration scheme for this method is:

$$x^{(r+1)} = \frac{1}{a_1}\left(d_1 - b_1y^{(r)} - c_1z^{(r)}\right)$$

$$y^{(r+1)} = \frac{1}{b_2}\left(d_2 - a_2x^{(r)} - c_2z^{(r)}\right)$$

$$z^{(r+1)} = \frac{1}{c_3}\left(d_3 - a_3x^{(r)} - b_3y^{(r)}\right)$$

The iteration is stopped when the values x, y, z start repeating with the desired degree of accuracy.

Gauss-Seidel Method

This method is only a refinement of Gauss-Jacobi method. In this method, once a new value for a unknown is found, it is used immediately for computing the new values of the unknowns.

If the rth iterates are $x^{(r+1)}, y^{(r+1)}, z^{(r+1)}$, then the iteration scheme for this method is:

$$x^{(r+1)} = \frac{1}{a_1}\left(d_1 - b_1 y^{(r)} - c_1 z^{(r)}\right)$$

$$y^{(r+1)} = \frac{1}{b_2}\left(d_2 - a_2 x^{(r+1)} - c_2 z^{(r)}\right)$$

$$z^{(r+1)} = \frac{1}{c_3}\left(d_3 - a_3 x^{(r+1)} - b_3 y^{(r+1)}\right)$$

Hence finding the values of the unknowns, we use the latest available values on the R.H.S. The process of iteration is continued until the convergence is obtained with desired accuracy.

Conditions for Convergence

Gauss-seidel method will converge if in each equation of the given system, the absolute values o the largest coefficient is greater than the absolute values of all the remaining coefficients.

$$|a_{ii}| > \sum_{j=1\ j\neq 1}^{n} |a_{ij}| \forall i = 1,2,3.........n$$

This is the sufficient condition for convergence of both Gauss-Jacobi and Gauss-seidel iteration methods.

Rate of Convergence

The rate of convergence of gauss-seidel method is roughly two times that of Gauss-Jacobi method. Further the convergences in Gauss-Seidel method is very fast in gauss-Jacobi.

Since the current values of the unknowns are used immediately in each stage of iteration for getting the values of the unknowns.

Advantages of iterative methods over direct methods for solving a system of linear equations:

Iterative Method	Direct Method
Approximate solution	Exact solution
Time consuming laborious	Less time
Converge only for special system	Any non-singular system

Problems

1. Let us solve the system of equations by Gauss-Jordan method: $5x_1 - x_2 = 9; - x_1 + 5x_2 - x_3 = 4; -x_2 + 5x_3 = -6.$

Solution:

$$(A,B) = \begin{pmatrix} 5 & -1 & 0 & 9 \\ -1 & 5 & -1 & 4 \\ 0 & -1 & 5 & -6 \end{pmatrix} R_1 \leftrightarrow R_2$$

$$\approx \begin{pmatrix} -1 & 5 & -1 & 4 \\ 5 & -1 & 0 & 9 \\ 0 & -1 & 5 & -6 \end{pmatrix} R_1 \leftrightarrow r_2 + 5R_1$$

$$\approx \begin{pmatrix} -1 & 5 & -1 & 4 \\ 0 & 24 & -5 & 29 \\ 0 & -1 & 5 & -6 \end{pmatrix} R_2 \leftrightarrow R_3$$

$$\approx \begin{pmatrix} -1 & 5 & -1 & 4 \\ 0 & -1 & 5 & -6 \\ 0 & 24 & -5 & 29 \end{pmatrix} \begin{matrix} R_1 \to R_1 + 5R_2 \\ R_3 \to R_3 + 24R_2 \end{matrix}$$

$$\approx \begin{pmatrix} -1 & 0 & 24 & -26 \\ 0 & -1 & 5 & -6 \\ 0 & 0 & 115 & -115 \end{pmatrix} R_3 \to R_3 / 115$$

$$\approx \begin{pmatrix} -1 & 0 & 24 & -26 \\ 0 & -1 & 5 & -6 \\ 0 & 0 & 1 & -1 \end{pmatrix} \begin{matrix} R_1 \to R_1 - 24R_3 \\ R_2 \to R_2 - 5R_3 \end{matrix}$$

$$\approx \begin{pmatrix} -1 & 0 & 0 & -2 \\ 0 & -1 & 0 & -1 \\ 0 & 0 & 1 & -1 \end{pmatrix} \begin{matrix} R_1 \to R_1 \times (-1) \\ R_2 \to R_2 \times (-1) \end{matrix}$$

$$(A,B) \approx \begin{pmatrix} 1 & 0 & 0 & 2 \\ 0 & 1 & 0 & 1 \\ 0 & 0 & 1 & -1 \end{pmatrix}$$

$x_1 = 2; \; x_2 = 1 \;\&\; x_3 = -1$

2. Apply Gauss-Seidel method to solve the system of equations $20x+y-2z=17$; $3x+2y-z=-18$; $2x-3y+20z=25$.

Solution:

The given system of equation satisfying diagonally dominant condition.

$$x^{(n+1)}=\frac{1}{20}\left[17-y^{(n)}+2z^{(n)}\right]$$

$$y^{(n+1)}=\frac{1}{20}\left[-18-3x^{(n+1)}+z^{(n)}\right]$$

$$z^{(n+1)}=\frac{1}{20}\left[25-2x^{(n+1)}+3y^{(n+1)}\right] \quad ...(1)$$

Seidel Iteration formula is:

Let the initial value be $y^{(0)}=0;\ z^{(0)}=0$

First Iteration:

Put,

n = 0 in (I),

$$x^{(1)}=\frac{17}{20}=0.85$$

$$y^{(1)}=\frac{1}{20}\left[-18-3(0.85)\right]=-1.028$$

$$z^{(1)}=\frac{1}{20}\left[25-2(0.85)+3(-1.028)\right]=1.011$$

Second Iteration:

$$x^{(2)}=\frac{1}{20}\left[17+1.028+2(1.011)\right]=1.003$$

$$y^{(2)}=\frac{1}{20}\left[-18-3(1.003)+1.011\right]=-1$$

$$z^{(2)}=\frac{1}{20}\left[25-2(1.003)+3(-1)\right]=1$$

Third Iteration:

$$x^{(3)}=\frac{1}{20}\left[17+1+2\right]=1$$

$$y^{(3)}=\frac{1}{20}\left[-18-3+1\right]=-1$$

$$z^{(3)} = \frac{1}{20}[25-2-3] = 1.$$

Hence from second and Third iteration the solution of system of equation is x = 1; y = - 1 and z = 1.

3. LetussolvebyGauss-Seidelmethod,thefollowingsystem: $28x+4y-z=32$, $x+3y+10z=24$, $2x+17y+4z=35$.

Solution:

Rearrange the equation in which it satisfying diagonally dominant condition:

$$28x+4y-z=32$$
$$2x+17y+4z=35$$
$$x+3y+10z=24$$

Gauss-Seidel iterative formula is:

$$x^{(n+1)} = \frac{1}{28}\left[32-4y^{(n)}+z^{(n)}\right]$$

$$y^{(n+1)} = \frac{1}{17}\left[35-2x^{(n+1)}+4z^{(n)}\right]$$

$$z^{(n+1)} = \frac{1}{10}\left[24-x^{(n+1)}-3y^{(n+1)}\right]$$

Let the initial approximation $y^{(0)} = z^{(0)} = 0$.

First Iteration:

$$x^{(1)} = \frac{1}{28}\left[32-4y^{(0)}+z^{(0)}\right] = \frac{32}{48} = 1.1429$$

$$y^{(1)} = \frac{1}{17}\left[35-2x^{(1)}+4z^{(0)}\right]$$

$$= \frac{1}{17}\left[35-2(1.1429)\right] = 1.9244$$

$$z^{(1)} = \frac{1}{10}\left[24-x^{(1)}-3y^{(1)}\right]$$

$$= \frac{1}{10}\left[24-1.1429-3(1.9244)\right] = 1.7084$$

Second Iteration:

$$x^{(2)} = \frac{1}{28}\left[32-4y^{(1)}+z^{(1)}\right] = 0.929$$

$$y^{(2)} = \frac{1}{17}\left[35 - 2x^{(2)} + 4z^{(1)}\right] = 1.5476$$

$$z^{(1)} = \frac{1}{10}\left[24 - x^{(2)} - 3y^{(2)}\right] = 1.8428.$$

Third Iteration:

$$x^{(3)} = 0.9876;\ y^{(3)} = 1.509;\ z^{(3)} = 1.8485$$

Fourth Iteration:

$$x^{(4)} = 0.9933;\ y^{(4)} = 1.507;\ z^{(4)} = 1.8486$$

$$x^{(5)} = 0.9936;\ y^{(5)} = 1.507;\ z^{(5)} = 1.8485$$

$$x^{(6)} = 0.9936;\ y^{(6)} = 1.507;\ z^{(6)} = 1.8485$$

Here fifth and sixth iteration values are same.

Hence x = 0.9936; y = 1.507; z = 1.8485.

4. Let us solve the following system of equations using Gauss-Seidel method:

$$10x + 2y + z = 9;\ x + 10y - z = -22;\ -2x + 3y + 10z = 22$$

Solution:

The given system of equation satisfies diagonally dominant condition:

Iterative formula is:

$$x^{(n+1)} = \frac{1}{10}\left[9 - 2y^{(n)} + z^{(n)}\right]$$

$$y^{(n+1)} = \frac{1}{10}\left[-22 - x^{(n+1)} + z^{(n)}\right]$$

$$z^{(n+1)} = \frac{1}{10}\left[22 + x^{(n+1)} - 3y^{(n+1)}\right]$$

Let the initial values of y = 0 and z = 0.

First iteration:

Put n = 0 in iterative formula:

$$x^{(1)} = \frac{1}{10}\left[9 - 2\,y^{(0)} + z^{(0)}\right] = \frac{9}{10} = 0.9$$

$$y^{(1)} = \frac{1}{10}\left[-22 - x^{(1)} + z^{(0)}\right] = \frac{1}{10}[-22 - 0.9]$$
$$= -2.29$$
$$z^{(1)} = \frac{1}{10}\left[22 + 2\,x^{(1)} - 3\,y^{(1)}\right]$$
$$= \frac{1}{10}\left[22 + 2\,x(0.9) - 3(-2.29)\right]$$
$$= 3.067$$

Iteration	$x^{(n+1)} = \frac{1}{10}\left[9 - 2y^{(n)} + z^{(n)}\right]$	$y^{(n+1)} = \frac{1}{10}\left[-22 - x^{(n+1)} + z^{(n)}\right]$	$z^{(n+1)} = \frac{1}{10}\left[22 + 2x^{(n+1)} - 3y^{(n+1)}\right]$
I	0.9	-2.29	3.067
II	1.0513	-1.9984	3.0098
III	0.9987	-1.9984	2.9994
IV	0.9998	-2.0	3.0
V	I	-2	3

From Fourth and Fifth iteration, solution of the given system is x = l, y =-2-and z =3.

5. Let us find all the eigen values and eigen vectors of the matrix $\begin{bmatrix} 1 & \sqrt{2} & 2 \\ \sqrt{2} & 3 & \sqrt{2} \\ 2 & \sqrt{2} & 1 \end{bmatrix}$ using Jacobi method.

Solution:

$$A = \begin{bmatrix} 1 & \sqrt{2} & 2 \\ \sqrt{2} & 3 & \sqrt{2} \\ 2 & \sqrt{2} & 1 \end{bmatrix}$$

Given:

Choose the largest off-diagonal element of the given matrix is 2 which are in the position a_{13}.

The suitable orthogonal matrix is:

$$p = \begin{bmatrix} \cos\theta & \theta & -\sin\theta \\ 0 & 1 & 0 \\ \sin\theta & 0 & \cos\theta \end{bmatrix}$$

$$\theta = \frac{1}{2}\tan^{-1}\left[\frac{2a_{13}}{a_{11} - a_{33}}\right] = \frac{1}{2}\tan^{-1}\left(\frac{4}{1-1}\right)$$

$$= \frac{1}{2}\tan(\alpha) = \frac{1}{2}\frac{\pi}{2} = \frac{\pi}{4}$$

$$\theta = \frac{\pi}{4}$$

$$\therefore \quad P = \begin{bmatrix} \frac{1}{\sqrt{2}} & 0 & \frac{1}{\sqrt{2}} \\ 0 & 1 & 0 \\ \frac{1}{\sqrt{2}} & 0 & \frac{1}{\sqrt{2}} \end{bmatrix}$$

The first transformation is, $D_1 = P^T AP$.

$$D_1 = \begin{bmatrix} \frac{1}{\sqrt{2}} & 0 & \frac{1}{\sqrt{2}} \\ 0 & 1 & 0 \\ -\frac{1}{\sqrt{2}} & 0 & \frac{1}{\sqrt{2}} \end{bmatrix} \begin{bmatrix} 1 & \sqrt{2} & 2 \\ \sqrt{2} & 3 & \sqrt{2} \\ 2 & \sqrt{2} & 1 \end{bmatrix} \begin{bmatrix} \frac{1}{\sqrt{2}} & 0 & -\frac{1}{\sqrt{2}} \\ 0 & 1 & 0 \\ \frac{1}{\sqrt{2}} & 0 & \frac{1}{\sqrt{2}} \end{bmatrix}$$

$$= \begin{bmatrix} \frac{1}{\sqrt{2}} & 0 & \frac{1}{\sqrt{2}} \\ 0 & 1 & 0 \\ -\frac{1}{\sqrt{2}} & 0 & \frac{1}{\sqrt{2}} \end{bmatrix} \begin{bmatrix} \frac{3}{\sqrt{2}} & \sqrt{2} & \frac{1}{\sqrt{2}} \\ 2 & 3 & 0 \\ \frac{3}{\sqrt{2}} & \sqrt{2} & -\frac{1}{\sqrt{2}} \end{bmatrix}$$

$$D_1 = \begin{bmatrix} 3 & 2 & 0 \\ 2 & 3 & 0 \\ 0 & 0 & -1 \end{bmatrix}$$

In D_1 the largest off diagonal dt is 2 is in the position a_{12}.

$$\therefore \quad Q = \begin{bmatrix} \cos\theta & -\sin\theta & 0 \\ \sin\theta & \cos\theta & 0 \\ 0 & 0 & 1 \end{bmatrix}$$

Where,

$$\theta = \frac{1}{2}\tan^{-1}\begin{bmatrix} 2 & a_{12} \\ a_{11} & -a_{22} \end{bmatrix} = \frac{1}{2}\tan^{-1}(\infty) = \frac{\pi}{4}$$

$$Q=\begin{bmatrix} \frac{1}{\sqrt{2}} & -\frac{1}{\sqrt{2}} & 0 \\ \frac{1}{\sqrt{2}} & \frac{1}{\sqrt{2}} & 0 \\ 0 & 0 & 1 \end{bmatrix}$$

$$D_2 = Q^T D_1 Q$$

$$D_2 = \begin{bmatrix} \frac{1}{\sqrt{2}} & -\frac{1}{\sqrt{2}} & 0 \\ -\frac{1}{\sqrt{2}} & \frac{1}{\sqrt{2}} & 0 \\ 0 & 0 & 1 \end{bmatrix} \begin{bmatrix} 3 & 2 & 0 \\ 2 & 3 & 0 \\ 0 & 0 & -1 \end{bmatrix} \begin{bmatrix} \frac{1}{\sqrt{2}} & 0 & 0 \\ \frac{1}{\sqrt{2}} & 1 & 0 \\ 0 & 0 & 1 \end{bmatrix}$$

$$= \begin{bmatrix} 5 & 0 & 0 \\ 0 & 1 & 0 \\ 0 & 0 & -1 \end{bmatrix} \text{ is a diagonal matrix.}$$

The eigen values are 5,1, - 1.

To find eigen vector:

The eigen vectors of A is PQ = (S).

$$PQ = \begin{bmatrix} \frac{1}{\sqrt{2}} & 0 & -\frac{1}{\sqrt{2}} \\ 0 & 1 & 0 \\ +\frac{1}{\sqrt{2}} & 0 & \frac{1}{\sqrt{2}} \end{bmatrix} \begin{bmatrix} \frac{1}{\sqrt{2}} & -\frac{1}{\sqrt{2}} & 0 \\ \frac{1}{\sqrt{2}} & \frac{1}{\sqrt{2}} & 0 \\ 0 & 0 & 1 \end{bmatrix}$$

$$= \begin{bmatrix} \frac{1}{2} & -\frac{1}{2} & -\frac{1}{\sqrt{2}} \\ \frac{1}{\sqrt{2}} & \frac{1}{\sqrt{2}} & 0 \\ +\frac{1}{2} & -\frac{1}{2} & \frac{1}{\sqrt{2}} \end{bmatrix}$$

Hence the eigen values are 5, 1, -1 and the corresponding eigenvectors are:

$$\begin{pmatrix} \frac{1}{2} \\ \frac{1}{\sqrt{2}} \\ \frac{1}{2} \end{pmatrix}, \begin{pmatrix} -\frac{1}{2} \\ \frac{1}{\sqrt{2}} \\ -\frac{1}{2} \end{pmatrix} \text{ and } \begin{pmatrix} \frac{1}{\sqrt{2}} \\ 0 \\ \frac{1}{\sqrt{2}} \end{pmatrix}$$

2.4 Numerical Differentiation and Numerical Integration

Numerical Differentiation

In the case of differentiation, we first write the interpolating formula on the interval (x_0, x_n) and the differentiate the polynomial term by term to get an approximated polynomial to the derivative of the function. When the tabular points are equidistant, one uses the Newton's Forward/ Backward Formula or Sterling's Formula; otherwise Lagrange's formula is used. Newton's Forward/ Backward formula is used depending upon the location of the point at which the derivative is to be computed. In case the given point is near the midpoint of the interval, Sterling's formula can be used. We illustrate the process by taking (i) Newton's Forward formula and (ii) Sterling's formula.

Recall, that the Newton's forward interpolating polynomial is given by:

$$f(x) = f(x_0 + h_u) \approx$$

$$y_0 + \Delta y_0 u + \frac{\Delta^2 y_0}{2!}(u(u-1)) + ... + \frac{\Delta^k y_0}{k!}\{u(u-1)....(u-k+1)\}$$
$$+ ... + \frac{\Delta^n y_0}{n!}\{u(u-1)...(u-n+1)\}$$

Differentiating above eq, we get the approximate value of the first derivative at x as:

$$\frac{df}{dx} = \frac{1}{h}\frac{df}{du} \approx \frac{1}{h}\begin{bmatrix} \Delta y_0 + \frac{\Delta^2 y_0}{2!}(2u-1) + \frac{\Delta^3 y_0}{3!}(3u^2 - 6u + 2) \\ + ... \frac{\Delta^n y_0}{n!}\left(nu^{n-1} - \frac{n(n-1)^2}{2}u^{n-2} + ... + (-1)^{(n-1)}(n-1)! \right) \end{bmatrix}$$

Where,

$$u = \frac{x - x_0}{h}$$

Thus, an approximation to the value of first derivative at x = x_o i.e. u = o is obtained as:

$$\left.\frac{df}{dx}\right|_{x=x_o} = \frac{1}{h}\left[\Delta y_o - \frac{\Delta^2 y_o}{2} + \frac{\Delta^3 y_o}{3} - .... + (-1)^{(n-1)}\frac{\Delta^n y_o}{n}\right]$$

Similarly, using Stirling's formula:

$$f(x_o^* + hu) \approx y_o^* + u\frac{\Delta y_{-1}^* + \Delta y_o^*}{2} + u^2\frac{\Delta^2 y_{-1}^*}{2!} + \frac{u(u^2-1)}{2} + \frac{\Delta^3 y_{-2}^* + \Delta^3 y^*_{-1}}{3!}$$

$$+u^2(u^2-1)\frac{\Delta^4 y_{-2}^*}{4!} + \frac{u(u^2-1)(u^2-2^2)}{2}\frac{\Delta^5 y_{-3}^* + \Delta^5 y_{-2}^*}{5!} + ...$$

Therefore,

$$\frac{df}{dx} = \frac{1}{h}\frac{df}{du} \approx \left[\begin{array}{l}\frac{\Delta y_{-1}^* + \Delta y_o^*}{2} + u\Delta^2 y_{-1}^* + \frac{(3u^2-1)}{2} \times \frac{(\Delta^3 y_{-2}^* + \Delta^3 y_{-1}^*)}{3!} \\ +2u(2u^2-1)\frac{\Delta^4 y_{-2}^*}{4!} + \frac{(5u^4-15u^2+4)(\Delta^5 y_{-3}^* + \Delta^5 y_{-2}^*)}{2\times 5!} + ...\end{array}\right]$$

Thus, the derivative at $x = x_o^*$ is obtained as:

$$\left.\frac{df}{dx}\right|_{x=x_o^n} = \frac{1}{h}\left[\frac{\Delta y_{-1}^* + \Delta y^*_o}{2} - \frac{(1)}{2} \times \frac{(\Delta^3 y_{-2}^* + \Delta^3 y_{-1}^*)}{3!} + \frac{4\times(\Delta^5 y_{-3}^* + \Delta^5 y_{-2}^*)}{2\times 5!} + ...\right]$$

Numerical Integration

Numerical Integration is the process of computing the value of a definite integral, $\int_a^b f(x)dx$, when the values of the integrand function, $y = f(x)$ are given at some tabular points. As in the case of Numerical differentiation, here also the integrand is first replaced with an interpolating polynomial, and then the integrating polynomial is integrated to compute the value of the definite integral. This gives us 'quadrature formula' for numerical integration. In the case of equidistant tabular points, either the Newton's formulae or Stirling's formula are used. Otherwise, one uses Lagrange's formula for the interpolating polynomial.

2.4.1 Trapezoidal rule, Simpson's 1/3rd and 3/8 Rule , Generalized Quadrature

The process of computing the value of a definite integral $\int_a^b y\,dx$ from a set of values

(x_i, y_i), i = 0,1,2,..... Where $x_o = a;\ x_n = b$ of the function y=f(x) is called Numerical integration.

Here the function y is replaced by an interpolation formula involving finite differences and then integrated between the limits a and b, the value $\int_a^b y\,dx$ is found.

General Quadrature formula for equidistant ordinates:

On simplification we obtain:

$$\int_{x_0}^{x_n} y\,dx = nh\left[y_o + \frac{n}{2}\Delta y_o + \frac{n(2n-3)}{12}\Delta^2 y_o + \frac{n(n-2)^2}{24}\Delta^3 y_o + \ldots\right]$$

This is the general Quadrature formula.

By putting n=1, Trapezoidal rule is obtained.

By putting n=2, Simpson's 1/3 rule is derived.

By putting n=3, Simpson's 3/8 rule is derived.

Problems

1. Let us evaluate $\int_0^1 \frac{dx}{1+x^2}$ using Trapezoidal rule.

Solution:

Take $n = 6 \Rightarrow h = \frac{1-0}{6} = \frac{1}{6}$

X	0	$\frac{1}{6}$	$\frac{2}{6}$	$\frac{3}{6}$	$\frac{4}{6}$	$\frac{5}{6}$	1
$y = \frac{1}{1+x_2}$	1	0.973	0,9	0.8	0.692	0.59	0.5
	y_o	y_1	y_2	y_3	y_4	y_5	y_6

$$\int_0^1 \frac{dx}{1+x^2} = \frac{h}{2}\left[(y_o + y_6) + (y_1 + y_2 + y_3 + y_4 + y_5)\right]$$

$$= \frac{1}{12}\left[(1+0.5) + 2(0.973 + 0.9 + 0.8 + 0.692 + 0.59)\right]$$

$$= 0.0833[9.41] = 0.7839$$

2. Let us evaluate $\int_{2}^{2.4}\int_{4}^{4.4} xy \ dx \ dy$ by Trapezoidal rule taking h = k = 0.1.

Solution:

Given:

$$h = k = 0:1.$$

$$h = \frac{x_n - x_0}{n} \Rightarrow 0.1 = \frac{4.4-4}{n} \Rightarrow n = 4$$

$$f(x,y) = xy$$

y / x	4	4.1	4.2	4.3	4.4
2.0	8.0	8.2	8.4	8.6	8.8
2.1	8.4	8.61	8.82	9.03	9.24
2.2	8.8	9.02	9.24	9.46	9.68
2.3	9.2	9.43	9.66	9.89	10.12
2.4	9.6	9.84	10.08	10.32	10.56

Trapezoidal Rule

$I = \int_{2}^{2.4}\int_{4}^{4.4} x \ y \ dx \ dy = \frac{hk}{4}$ [(Sum of values of f at the four corners)

+ 2 (Sum of values of f at the remaining nodes on the boundary)

+ 4 (Sum of values of f at the interior nodes)].

$$= \frac{(0.1)(0.1)}{4}[(8 + 8.8 + 10.56 + 9.6)$$

$$+2(8.2+8.4+8.6+9.24+9.68+10.12+10.32+10.08+9.84+9.2+8.8+8.4)$$

$$+ 4(8.61+8.82+9.03+9.02+9.24+9.46+9.43+9.66+9.89)]$$

$$I = (0.0025)[36.96+2(110.88)+4(83.16)]$$

$$= 1.4784$$

3. Let us evaluate $\int_{0}^{1}\int_{0}^{1} \frac{1}{1+x+y} dx \, dy$ by trapezoidal rule.

Solution:

Take,

$$h = k = 0.25$$

Here,

$$f(x, y) = \frac{1}{1+x+y}$$

x	0	0.25	0.5	0.75	1
0	1	0.8	0.6667	0.5714	0.5
0.25	0.8	0.6667	0.5714	0.5	0.4444
0.5	0.667	0.5714	0.5	0.4444	0.4
0.75	0.5714	0.5	0.4444	0.4	0.3636
1	0.5	0.4444	0.4	0.3636	0.33

$$= \int_0^1 \int_0^1 \frac{dx\,dy}{1+x+y} = \frac{hk}{4}$$ [(Sum of Values of f at the four comers)

+ 2(Sum of the values off at the remaining nodes on the boundary)

+ 4(Sum of the values of f at the interior nodes)]

$$\begin{aligned} I = \frac{(0.25)(0.25)}{4}&[(1+0.5+0.3333+0.5) \\ &+2(0.8+0.6667+0.5714+0.4444+0.4+0.3636 \\ &+0.3636+0.4+0.4444+0.5714+0.6667+0.8) \\ &+4(0.6667+0.5714+0.5+0.5714+0.5 \\ &+04444+0.5+0.4444+0.4)] \end{aligned}$$

$$\begin{aligned} I &= (0.015625)\left[2.3333+2(6.4923)+4(4.5983)\right] \\ &= 0.5267 \end{aligned}$$

4. Using Trapezoidal rule, let us evaluate $\int_1^2 \int_1^2 \frac{dx\,dy}{x^2+y^2}$ numerically with h = 0.2 along x-direction and k = 0.25 along y-direction.

Solution:

Xy	1	1.2	1.4	1.6	1.8	2
1	0.5	0.4098	0.3378	0.2809	0.2359	0.2

1.25	0.3902	0.3331	0.2839	0.2426	0:2082	0.1798
1.5	0.3077	0.2710	0.2375	0.2079	0.1821	0.16
1.75	0.2462	0.2221	0.1991-	0.1779	0.1587	0.1416
2	0.02	0.1838	0.1679	0.1524	0.1381	0.125

Trapezoidal Rule

$$\int_c^d \int_a^b f(x,y)dx\ dy$$

$$=\frac{hk}{4}\text{[(Sum of Values of f at the four corners)}$$

+ 2 (Sum of the Values of f at the remaining nodes on the boundary)

+ 4 (Sum of the Values of f at the interior nodes)]

$$\therefore \int_1^2 \int_1^2 \frac{1}{x^2+y^2} dx\ dy = \frac{(0.2)(0.25)}{4}$$

$$\begin{aligned}
&[(0.5+0.2+0.125+0.2)\\
&+2(0.4098+0.3378+0.2809+0.2359+0.1798+0.16\\
&+0.416+0.1381+0.1524+0.1679+0.1838+0.2462\\
&+0.3077+0.3902)\\
&+4(0.3331+0.2839+0.2426+0.2082+0.2710+0.2375]\\
&+0.2079+08121+0.2221+0.1991+0.1779+0.1587)]\\
&=(0.0125)[1.025+6.6566+13.4625]\\
&=0.2643
\end{aligned}$$

Simpson's One-third Rule

Let y = f(x) be a function which takes the value y_0, y_1, y_2, ..., y_n corresponding to x_0, x_1, x_2,..., x_n then,

$$\int_{x_0}^{x_n} y\ dx \frac{h}{3}\left[(y_0+y_n)+2(y_2+y_4+...)+4(y_1+y_3+...)\right]$$

Where,

$$h=\frac{x_n-x_0}{n};\ \text{n is even.}$$

The Local Error Term in Simpson's $\frac{1}{3}$ Rule

The principal part of the error in (x_{0}, x_2) is $-\frac{h^2}{90}y_0^{IV}$, where y_0^{IV} is the value of fourth derivative of y at $x = x_0$.

5. Let us evaluate $I = \int_1^{1.4}\int_2^{2.4}\frac{1}{xy}dx\ dy$ using Simpson's one-third rule.

Solution:

Let,

$$I = \int_{y_0}^{y_n}\int_{x_0}^{x_n} f(x,y)dx\ dy$$

Here,

$$x_0 = 2;\ x_n = 2.4;\ y_0 = 1;\ y_n = 1.4$$

$$h = \frac{2.4-2}{4} = 0.1\ \&\ k = \frac{1.4-1}{4} = 0.1$$

$$f(x,y) = \frac{1}{xy}$$

Divide the range of x and y into 4 equal parts.

x / y	2	2.1	2.2	2.3	2.4
1	0.5	0.4762	0.4545	0.4348	0.4167
1.1	0.4545	0.4329	0.4132	0.3953	0.3788
1.2	0.4167	0.3968	0.3788	0.3623	0.3472
1.3	0.3846	0.3663	0.3497	0.3344	0.3205
1.4	0.3571	0.3401	0.3247	0.3106	0.2976

Simpson's one-third rule is:

$I = \frac{hk}{9}$ [(Sum of the values of f at the four corners).

+ 2 (Sum of the values of f at the odd position on the boundary except the corners).

+ 4 (Sum of the values of f at the even position on the boundary except the corners).

+ {4(Sum of the values of f at odd positions).

+ 8(Sum of the values of f at even position) on the odd row of the matrix except boundary row}.

+ {8(Sum of the values of f at odd positions).

+ 16(Sum of the values of f at even positions) on the even row of the table except boundary rows).

$$I = \frac{(9.1)(0.1)}{9}[(0.5+0.4167+0.3571+0.2976)$$
$$+2(0.4167+0.4545+0.3472+0.3247)$$
$$+4(0.3846+0.4545+0.4762+0.4348$$
$$+0.3788+0.3205+0.3106+0.3401)]$$
$$+[4(0.3788)+8(0.3968+0.3623)]$$
$$+[8(0.3497+0.4132)+16(0.3663+0.3344+0.4329+0.3953]$$

$$I = 0.0613$$

6. The velocities of a ear running on a straight road at intervals of 2 minutes are given below:

Time (min)	0	0	2	4	6	8	10	12
Velocity (km/hr)	0	22	30	27		18	7	0

Using Simpson's $\frac{1}{3}$rd rule, let us find the distance covered by the car.

Solution:

Given: $h = 2$

$$\text{Velocity} = \frac{ds}{dt} \Rightarrow ds = v\ dt$$

Distance covered by the car:

$$S = \int_0^{12} v\ dt$$

$$S = \int_0^{12} v\ dt = \frac{h}{3}\left[(y_0 + y_6) + 2(y_2 + y_4) + 4(y_1 + y_3)\right]$$

$$= \frac{2}{3}\left[(0+0) + 2(30+18) + 4(22+7)\right]$$

$$= 141.33 \text{ km/hr.}$$

7. Evaluate $I=\int_0^6 \frac{1}{1+x}dx$ by using (i) direct integration Trapezoidal rule (iii) Simpson's one third rule (iv) Simpson's three-eighth rule.

Solution:

Taking n = 6

$$h=\frac{\text{Upper Limit}-\text{Lower Limit}}{n}=\frac{6-0}{6}=-1$$

$$h=1$$

Here,

$$y=\frac{1}{1+x}$$

X	X_0	X_1	X_2	X_3	X_4	X_5	X_6
	0	1	2	3	4	5	6
$y=\frac{1}{1+x}$	1	0.5	0.3333	0.25	0.2	0.1667	0.1429
	y_0	y_1	y_2	y_3	y_4	y_5	y_6

Case (i): By direct integration:

$$I=\int_0^6 \frac{1}{14x}dx=\left[\log(1+x)\right]_0^6=\log 7-\log 1=1.946$$

Case (ii): By Trapezoidal rule:

$$I=\int_0^6 \frac{1}{14x}dx=\frac{h}{2}\left[(y_0+y_6)+2(y_1+y_2+y_3+y_4+y_5)\right]$$

$$=\frac{1}{2}\left[(1+0.1429)+2(0.5+0.3333+0.25+0.2+0.1667)\right]$$

$$=\frac{1}{2}[4.0429]=2.0215$$

Case (iii): By Simpson's one-third rule:

$$I=\int_0^6 \frac{1}{1+x}dx=\frac{h}{3}\left[(y_0+y_6)+2(y_2+y_4)+4(y_1+y_3+y_5)\right]$$

$$=\frac{1}{3}\left[(1+0.1429)+2(0.3333+0.2)+4(0.5+0.25+0.1667)\right]$$

$$=\frac{1}{3}[1.1429+1.0666+3.6668]$$

$$=\frac{1}{3}[5.8763]=1.9588$$

Case (iv): By Simpson's three-eight rule:

$$I=\int_{0}^{6}\frac{1}{1+x}dx=\frac{3h}{8}\left[(y_0+y_6)+3(y_1+y_2+y_4+y_5)+2(y_3)\right]$$

$$=\frac{3}{6}\left[(1+0.1429)+3(0.5+0.3333+0.2+0.1667)+2(0.25)\right]$$

$$=\frac{3}{8}[1.429+3.6+0.5]=1.9661$$

8. Compute $\int_{0}^{\pi/2} 1 \sin x\, dx$ using Simpson's 3/8 rule.

Solution:

To use Simpson's $\frac{3}{8}$ rule the interval should be a multiple of 3.

Consider n = 6.

$$h=\frac{\pi/2-0}{60}=\frac{\pi}{12}\ ;\ y=\sin x$$

X	0	$\frac{\pi}{12}$	$\frac{2\pi}{12}$	$\frac{3\pi}{12}$	$\frac{4\pi}{12}$	$\frac{5\pi}{12}$	$\frac{\pi}{2}$
y = sin x	0	0.2588	0.5	0.7071	0.866	0.9659	1
	y_0	y_1	y_2	y_3	y_4	y_5	$y_6=y_n$

Simpson's $\frac{3}{8}$ rule is:

$$\int_{0}^{\pi/2} y\ dx=\frac{3h}{8}\left[(y_0+y_6)+3(y_1+y_2+y_4+y_5)+2(y_3)\right]$$

$$\int_{0}^{\pi/2} \sin x\ dx=\frac{3\pi}{96}\left[(0+1)+3(0.2588+0.5+0.866+0.9659)+2(0.7071)\right]$$

$$=(0.0982)[10.1863]$$

$$=1.0003$$

9. Evaluate $\int_0^2 \int_0^1 4xy \; dx \, dy$ using Simpson's rule by taking $h = \frac{1}{4}$ and $k = \frac{1}{2}$.

Solution:

Given:

$$h = 1/4 \; \& \; k = 1/2. \; 0 < x < 1 \; \& 0 < y < 2.$$

Let,

$$I = \int_0^2 \int_0^1 4xy \; dx \, dy$$

Here,

$$f(x, y) = 4\,xy;$$

Simpson's rule is:

$I = \frac{hk}{9}$ (Sum of values of f at the four corners)

+ 2(Sum of the values of f at the odd positions on the boundary except the corner)

+ 4(Sum of the values of f at the even positions on the boundary)

+ {4(Sum of values of f at odd positions)

+ 8(Sum of values of at even positions)

On the odd row of the table except boundary row}

+ {8(Sum of values of f at odd positions)

+ 16(Sum of values of f at even positions)

On the even row except boundary row}]

x / y	0	0	0.5	1	1.5
0	0	0	0	0	0
0.25	0	0.5	1	1.5	2
0.5	0	1	2	3	4
0.75	0	1.5	3	4.5	6
1	0	2	4	6	8

$$I=\int_0^2\int_0^1 4\,xy\,dx\,dy=\frac{(0.25)(0.5)}{9}[(0+0+0+8)+2(4+4)$$

$$+4(2+6+2+6)+4(2)+8(1+3)+8(1+3)+16(0.5+1.5+4.5)]$$

$$=\frac{36}{9}=4$$

Romberg's Method

The Romberg's integration formula with and h_1 and h_2 is:

$$I=\frac{I_1h_2^2-I_2h_1^2}{h_2^2-h_1^2}$$

When h_1 = h and h_2 = h/2.

$$I=\frac{I_1\left(\frac{h^2}{4}\right)-I_2h^2}{\frac{h^2}{4}-h^2}$$

$$I=I_2+\frac{1}{3}(I_2-I_1)$$

10. Evaluate $\int_0^{\frac{1}{2}}\frac{x}{\sin x}dx$ correct to three decimal places using Romberg's method.

Solution:

Let,

$$y=\frac{x}{\sin x}\ \&\ 1=\int_0^{1/2}y\,dx$$

Case (i): Take $h=\frac{b-a}{2}=\frac{\left(\frac{1}{2}\right)-0}{2}=\frac{1}{4}=0.25$

x	0	0.25	0.5
$y=\frac{x}{\sin x}$	0	1.0105	1.0429

Using Trapezoidal rule is:

$$I_1 = \int_0^{1/2} \frac{x}{\sin x}\, dx = \frac{h}{2}\left[(y_0 + y_2) + 2y_1\right]$$

$$= \frac{1}{8}\left[(0 + 1.0429) + 2(1.0105)\right]$$

$$I_1 = 0.383$$

Case (ii): Take $h = \frac{1}{8} = 0.125$

x	0	0.125	0.25	0.375	0.5
$y = \frac{x}{\sin x}$	0	1.0026	1.0105	1.0238	1.0429
	y_0	y_1	y_2	y_3	y_4

$$I_2 = \frac{h}{2}\left[(y_0 + y_4) + 2(y_1 + y_2 + y_3)\right]$$

$$= \frac{1}{16}\left[(0 + 1.0429) + 2(1.0026 + 1.0105 + 1.0238)\right]$$

$$= 0.445$$

Case (iii): Take $h = \frac{1}{16} = 0.0625$

x	0	0.0625	0.125	0.1875	0.25	0.3125
y	0	1.0007	1.0026	1.0059	1.0105	1.0165
	y_0	y_1	y_2	y_3	y_4	y_5

X	0.375	0.4375	0.5
y	1.0238	1.0326	1.0429
	y_6	y_7	y_8

$$I_3 = \frac{h}{2}\left[(y_0 + y_8) + 2(y_1 + y_2 + y_3 + y_4 + y_5 + y_6 + y_7)\right]$$

$$= 0.476$$

By Romberg's formula for I_1 & I_2 we have:

$$I = I_2 + \left(\frac{I_2 - I_1}{3}\right) = 0.445 + \frac{(0.445 - 0.383)}{3}$$

$$= 0.4657$$

By Romberg's formula for I_2 & I_3 we have:

$$I = I_3 + \left(\frac{I_3 - I_2}{3}\right) = 0.476 + \frac{(0.476 - 0.445)}{3}$$

$$= 0.4863$$

Thus see Trapezoidal rule, evaluate $\int_{-1}^{1} \frac{1}{(1+x^2)}\, dx$ by taking eight equal intervals:

Let,

$$I = \int_{-1}^{1} \frac{1}{(1+x^2)}\, dx = \int_{a}^{b} y\ dx$$

Here,

$$a = -1; b = 1, n = 8;\ y = \frac{1}{1+x^2}$$

$$h = \frac{b-a}{n} = \frac{2}{8} = 0.25$$

X	-1	-0.75	-0.5	-0.25	0	0.25	0.5	0.75	1
$y = \frac{1}{1+x^2}$	0.5	0.64	0.8	0.9412	1	0.9412	0.8	0.64	0.5
	y_0	y_1	y_2	y_3	y_4	y_5	y_6	y_7	y_8

Trapezoidal rule is:

$$I = \int_{-1}^{1} \frac{1}{1+x^2}\, dx = \frac{h}{2}\left[(y_0 + y_8) + 2(y_1 + y_2 + y_3 + y_4 + y_5 + y_6 + y_7)\right]$$

$$= \frac{0.25}{2}\left[(0.5+0.5) + 2(0.64+0.8+0.9412+1+0.9412+0.8+0.64)\right]$$

$$I = 1.5656$$

11. Taking h = 0.05, let us evaluate $\int_{1}^{1.3} \sqrt{x}\ dx$ using Trapezoidal rule and Simpson's three-eighth rule.

Solution:

Given:

$$I = \int_{1}^{1.3} \sqrt{x}\, dx$$

Here,

$$y = \sqrt{x}$$

$$h = 0.05$$

x	1	1.05	11	1.15	1.2	1.25	1.3
$y = \sqrt{x}$	1	1.0247	1.0488	1.0724	1.0954	1.118	1.1402
	y_0	y_1	y_2	y_3	y_4	y_5	y_6

Trapezoidal Rule

$$I = \frac{h}{2}\left[(y_0 + y_6) + 2(y_1 + y_2 + y_3 + y_4 + y_5)\right]$$

$$= \frac{0.05}{2}\left[(1 + 1.1402) + 2(1.0247 + 1.0488 + 1.0724 + 1.0954 + 1.118)\right]$$

$$= 0.3215$$

Simpson's Three-Eighth Rule:

$$I = \frac{3h}{8}\left[(y_0 + y_6) + 3(y_1 + y_2 + y_4 + y_5) + 2y_3\right]$$

$$= \frac{3(0.05)}{8}\left[(1 + 1.402) + 3(1.0247 + 1.0488 + 1.0954 + 1.178) + 2(1.0724)\right]$$

$$I = 0.3215$$

Formulas:

- Newton's forward formula to find the derivatives is:

$$\frac{dy}{dx} = \frac{1}{h}\left[\Delta y_0 + \frac{2u-1}{2}\Delta^2 y_0 + \frac{3u^2 - 6u + 2}{6}\Delta^3 y_0 + \ldots\right]$$

$$\frac{d^2y}{dx^2} = \frac{1}{h^2}\left[\Delta^2 y_0 + (u-1)\Delta^3 y_0 + \frac{6u^2 - 18u + 11}{12}\Delta^4 y_0 + \ldots\right]$$

$$\frac{d^3y}{dx^3} = \frac{1}{h^3}\left[\Delta^3 y_0 + \frac{(12u - 18)}{12}\Delta^4 y_0 + \ldots\right] \text{ where } u = \frac{x - x_0}{h}$$

- Newton's forward formula to find the derivatives at $x = x_0$ is:

$$\left(\frac{dy}{dx}\right)_{x=x_0} = \frac{1}{h}\left[\Delta y_0 - \frac{\Delta^2 y_0}{2} + \frac{\Delta^3 y_0}{3} - \ldots\right]$$

$$\left(\frac{d^2y}{dx^2}\right)_{x=x_0}=\frac{1}{h^2}\left[\Delta^2y_0+\Delta^3y_0+\frac{11}{12}\Delta^4y_0+...\right]$$

$$\left(\frac{d^3y}{dx^3}\right)_{x=x_0}=\frac{1}{h^3}\left[\Delta^3y_0+\frac{3}{2}\Delta^4y_0+...\right]$$

- Newton's backward formula to find the derivatives is:

$$\frac{dy}{dx}=\frac{1}{h}\left[\nabla y_n-\frac{2v+1}{2}\nabla^2y_n+\frac{3v^2+6v+2}{6}\nabla^3y_n+...\right]$$

$$\frac{d^2y}{dx^2}=\frac{1}{h^2}\left[\Delta^2y_n+(v+1)\Delta^3y_n+\frac{6v^2+18v+11}{12}\Delta^4y_n+...\right]$$

$$\frac{d^3y}{dx^3}=\frac{1}{h^3}\left[\Delta^3y_n+\frac{12+18}{12}\Delta^4y_n+...\right]\text{ where }v=\frac{x-x_n}{h}$$

- Newton's backward formula to find the derivatives at $x = x_n$ is:

$$\left(\frac{dy}{dx}\right)_{X=X_n}=\frac{1}{h}\left[\nabla y_n+\frac{\nabla^2y_n}{2}+\frac{\nabla^3y_n}{3}+...\right]$$

$$\left(\frac{d^2y}{dx^2}\right)_{X=Xn}=\frac{1}{h^2}\left[\Delta^2y_n+\Delta^3y_n+\frac{11}{12}\Delta^4y_n+...\right]$$

$$\left(\frac{d^3y}{dx^3}\right)_{X=Xn}=\frac{1}{h^3}\left[\Delta^3y_n+\frac{3}{2}\Delta^4y_n+...\right]$$

- Trapezoidal Rule:

$$\int_{x_0}^{x_0+nh}f(x)dx=\frac{h}{2}\left[(y_0+y_n)+2(y_1+y_2+y_3+...)\right]$$

- Simpson's 1/3rd Rule:

$$\int_{x_0}^{x_0+nh}f(x)dx=\frac{h}{3}\left[(y_0+y_n)+4(y_1+y_3+y_5+...)+2(y_2+y_4+y_6+...)\right]$$

- Simpson's 3/8thRule:

$$\int_{x_0}^{x_0+nh}f(x)dx=\frac{3h}{3}\left[(y_0+y_n)+3(y_1+y_2+y_4+y_5+y_7+...)+2(y_3+y_6+y_9+...)\right]$$

- Romberg's formula for I_1 and I_2 is $I=I_2+\left(\frac{I_2-I_1}{3}\right)$

- Error is the Trapezoidal formula is of the order h^2 and in the Simpson formula is of the order h^4.

- Two points Gaussian Quadrature formula is:

$$\int_{-1}^{1} f(x)dx = f\left(\frac{-1}{\sqrt{3}}\right) + f\left(\frac{1}{\sqrt{3}}\right)$$

- Three points Gaussian Quadrature formula is:

$$\int_{-1}^{1} f(x)dx = \frac{5}{9}\left[f\sqrt{\left(-\frac{3}{5}\right)} + f\sqrt{\left(\frac{3}{5}\right)}\right] + \frac{8}{9}f(0)$$

- If the range is not (-1,1) then the idea to solve the Gaussian Quadrature problem is:

$$x = \left(\frac{b-a}{2}\right)z + \left(\frac{b+a}{2}\right)$$

2.5 Solution by Taylor's Series Method

Consider the first order differential equation $\frac{dy}{dx} = f(x,y), y(x_0) = y_0$.

The solution of the above initial value problem is obtained in two types as:

- Power series solution.
- Point wise solution.

(i) Power series solution:

$$y(x) = y_0 + \frac{x-x_0}{1!}y_0' + \frac{(x-x_0)^2}{2!}y_0'' + \frac{(x-x_0)^3}{3!}y_0''' + \ldots..$$

Where,

$$y_m^{(r)} = \frac{d^r y}{dx^r} \text{ at } (x_0, y_0)$$

(ii) Point wise solution:

$$y_{m+1} = y_m + \frac{h}{1!}y_m' + \frac{h^2}{2!}y_m'' + \frac{h^3}{2!}y_m''' + \ldots$$

Where,

$$y_m^{(r)} = \frac{d^r}{dx^r} \text{at} \left(x_m, y_m\right)$$

and m=0,1,2.......

Problems

1. Let us find y(1.1) if y = x + y, y(l) = 0 by Taylor series method.

Solution:

Given:

$$\begin{aligned}
&y' = x + y;\ x_0 = 1;\ y_0 = 0;\ x_1 = 1.1 \,\&\, h = 0.1 \\
&y' = x + y;\ y'_0 = x_0 + y_0 = 1 \\
&y'' = 1 + y';\ y''_0 = 1 + y'_0 = 2 \\
&y''' = y'';\ y'''_0 = y''_0 = 2.
\end{aligned}$$

Taylor series:

$$\begin{aligned}
y(1.1) &= y_0 + \frac{h}{1!} y_0 + \frac{h^2}{2!} y_0 + \frac{h^3}{3!} y''_0 \\
&= 0 + (0.1)(1) + \frac{(0.1)^2 (2)}{(2)} + \frac{(0.1)^3 (2)}{6} + \ldots \\
y(1.1) &= 0.1103.
\end{aligned}$$

2. Let us find y(0.1) if $\frac{dy}{dx} = 1 + y, y(0) = 1$ using Taylor series method.

Solution:

Given:

$$y' = 1 + y;\ x_0 = 0;\ y_0 = 1;\ x_1 = 0.1;\ h = 0.1.$$

$$y_1 = y_0 + \frac{h}{1!} y'_0 + \frac{h^2}{2!} y''_0 + \cdots \qquad \ldots(1)$$

$$\begin{aligned}
&y' = 1 + y;\ y'_0 = 1 + y_0 = 2 \\
&y'' = y';\ y''_0 = y'_0 = 2 \\
&y''' = y'';\ y'''_0 = y''_0 = 2
\end{aligned}$$

$$\therefore\ y_1 = 1 + (0.1)(2) + \frac{(0.1)^2(2)}{2} + \frac{(0.1)^3(2)}{6} + \ldots$$

$$y(0,1) = 1 + 0.2 + 0.01 + 0.0003$$
$$= 1.2103$$

3. Using Taylor series method, let us find y(0.1) if $y = x^2 + y^2$, $y(0) = 1$.

Solution:

Given:

$$y' = x^2 + y^2;\ x_0 = 0;\ y_0 = 1;\ x = 0.1$$

$$h = x_{1-}\, x_0 = 0.1$$

Taylor series formula is:

$$y(x) = y_0 + \frac{h}{1!} y_0 + \frac{h^2}{2!} y_0 + \frac{h^3}{3!} y_0 + \ldots$$
$$y = x^2 + y^2;\ \ y_0 = x_0^2 + y_0^2 = 0 + 1 = 1$$
$$y = 2x + 2yy;\ \ y_0 = 2x_0 + 2y_0 y_0 = 0 + 2(1)(1) = 2$$
$$y = 2 + \left[yy + (y)^2\right];\ \ y_0 = 8$$

$$y(0,1) = 1 + \frac{(0.1)}{1}(1) + \frac{(0.1)^2}{2}(2) + \frac{(0.1)^3}{6}(8)$$
$$= 1 + 0.1 + 0.01 + 0.0013$$
$$y(0,1) = 1.1113$$

4. Using Taylor's series method, let us find y at x = 1.1 by solving the equation $\frac{dy}{dx} = x^2 + y^2 : y(1) = 2$. Carryout the computations upto fourth order derivative.

Solution:

Given:

$$y' = x^2 + y^2;\ y(1) = 2$$

$$x_0 = 1;\ y_0 = 2.$$

To find:

$$y \text{ at } x = 1.1$$

$$\therefore x_1 = 1.1;\ h = 0.1$$

$$y' = x^2 + y^2;\ y'_0 = x_0^2 + y_0^2 = 1 + 4 = 5$$

$$y'' = 2x + 2yy';\ y''_0 = 2x_0 + 2y_0 y'_0 = 2 + 2(2)(5) = 22$$

$$y''' = 2 + 2[yy'' + y'_2];\ y'''_0 = 2 + 2[y_0 y''_0 + y'_0 2]$$

$$= 2 + 2\left[(2)\ (22).+\ 5^2\right]$$

$$= 140$$

$$y'v = 2[yy''' + y''y' + 2y'y''];\ y'ov = 2[yoy'''o + 3\,y'o\,y''o]$$

$$= 2[(2)(140) + 3(5)(22)] = 1220$$

Taylor Series is:

$$y(x) = y_0 + \frac{h}{1!}y'_0 + \frac{h^2}{2!}y''_0 + \frac{h^3}{3!}y'''_0 + \frac{h^4}{4!}y_0^{iv} + \ldots$$

$$y(1.1) = 2 + (0.1)(5) + \frac{(0.1)^2}{2}(22) + \frac{(0.1)^3}{6}(140) + \frac{(0.1)^4}{24}(1220)$$

$$= 2 + 0.5 + 0.11 + 0.0233 + 0.0051$$

$$y(1.1) = 2.6384$$

5. Let us find the value of y at x = 0.1, 0.2 given $\frac{dy}{dx} = x^2y - 1,\ y(0) = 1$, by Taylor's series method up to four terms.

Solution:

Given:

$$y' = x^2\,y - 1;\ x_0 = 0;\ y_0 = 1;\ x_1 = 0.1$$

Taylor Series Formula is:

$$y_1 = y_0 + \frac{h}{1!}y'_0 + \frac{h^2}{2!}y''_0 + \frac{h^3}{3!}y'''_0 + \frac{h^4}{4!}y_0^{iv} + \ldots$$

$y' = x^2y - 1 = -1$	$y'_0 = x_0^2\ y_0 - 1 = -1$
$y'' = 2xy\ +\ x^2y'$	$y''_0 = 2x_0y_0 + x_0^2y'_0 = 0$

$y''_0 = 2y + 4xy' + x^2y''$	$y''_0 = 2$
$y'^{v} = 6y'' + 6xy'' + x^2 6'''$	$y'^{v}_0 = -6$

$$h = x_1 - x_0 = 0.1$$

$$y_{(x_1)} = 1 + (0.1)(-1) + \frac{(0.1)^2}{2}(0) + \frac{(0.1)^3}{6}(2) + \frac{(0.1)^4}{24}(\cdots 6) + \cdots$$

$$y = (0.1) = 0.900305$$

To find y(0.2) here h = 0.2.

$$y(0.2) = 1 + (0.2)(-1) + \frac{(0.2)^2}{2}(0) + \frac{(0.2)^3}{6}(2) + \frac{(0.2)^4(-6)}{24} + \cdots$$

$$y(0.2) = 0.802266$$

2.5.1 Picard's Method of Successive Approximation

For $x \in I$ with $|x - x_a| \leq a$, define inductively:

$$y_0(x) = y_0 \text{ and for } n = 1,2,3\ldots,$$

$$y_n(x) = y_0 + \int_{z0}^{2} f(s, y_{n-1})(s)ds.$$

Then $y_0, y_1, \ldots, y_n, \ldots$ are known as Picard's successive approximations.

Problem

1. Let us find the Picard's successive approximations for the IVP.

$$y = f(y), 0 \leq x \leq 1, y \geq 0 \text{ and } y(0) = 0;$$

Where,

$$f(y) = \sqrt{y} \text{ for } y \geq 0$$

Solution:

By definition $y_0(x) = y_0 = 0$

And,

$$y_1(x) = y_0 + \int_0^x f(y_0)ds = 0 + \int_0^x \sqrt{0}\, ds = 0$$

A similar argument implies that $y_n(x) \equiv 0$ for all $n = 2,3,...$ and $\lim_{n\to\infty} y_n(x) \equiv 0$. Also, it will be easily verified that $y(x) \equiv 0$ is a solution of the IVP. Also $y(x) = \frac{x^2}{4}, \leq x \leq 1$ is a solution of eq. and the $\{y_n\}$'s do not converge to $\frac{x^2}{4}$. Note here that the IVP has at least two solutions.

2.6 Single Step Methods, Euler's Method and Euler's Modified Method

Introduction

The solution to a differential equation is to function that satisfies the differential equation and that also satisfies certain initial conditions on the function.

The solution of this differential equation involves n constants and these constants are determined with help of n conditions.

Single Step Methods

In these methods we use information about the curve at one point and we do not iterate the solution. The method involves more evaluation of the function.

We will discuss the Numerical solution by Taylor series method, Euler methods and Runge - Kutta methods all require the information at a single point $x = x_0$.

Euler's Method is a form of numerical integration that a way to approximate the solution of a first-order differential equation where the initial point on the solution curve is known, but the shape of the curve is unknown.

Problems

1. Let us find y(0, 1) by using Euler's method given that $\frac{dy}{dx} = x + y,\ y(0) = 1.$

Solution:

Given:

$$f(x,y) = x + y; x_0 = 0;$$

$$y_0 = 1; x_1 = 0.1; h = 0.1.$$

Euler's formula is:

$$\begin{aligned} y_1 &= y_0 + h\,f(x_0, y_0) \\ &= 1 + (0.1)\,f(0.1) \\ &= 1 + (0.1)(1) \end{aligned}$$

$y(0, 1) = 1.1$

2. Using Euler's method, let us find the solution of the initial value problem $y' = y - x^2 + 1$, $y(0) = 0.5$ at $x = 0.2$ taking $h = 0.2$.

Given:

$$y' = y - x^2 + 1.$$
$$x_0 = 0;\ y_0 = 0.5,\ h = 0.2;\ x_1 = 0.2$$

Solution:

Euler's formula is,

$$y_{n+1} = y_n + h\,f(x_n, y_n);\ n = 0, 1, 2,$$
$$y_1 = y_0 + h\,f(x_0, y_0)$$
$$= 0.5 + (0.2)\left[y_0 - x_0^2 + 1\right]$$
$$= 0.5 + (0.2)[0.5 - 0 + 1]$$

$y(0, 2) = 0.8$

3. Using Modified Euler's method, let us find y(4.1) and y(4.2) if $5x\frac{dy}{dx} = y^2 - 2 = 0$; y(4)=1.

Given:

$$5\,xy + y^2 - 2 = 0$$

$$\Rightarrow y = \frac{2 - y^2}{5x};\ x_0 = 4,\ y_0 = 1$$
$$x_1 = 4.1;\ x_2 = 4.2$$

To find:

y_1 and y_2.

Solution:

Here,

$$f(x,y) = \frac{2 - y^2}{5x},\ h = 0.1$$

Modified Euler's formula is:

$$y_{n+1} = y_n + hf\left[x_n + \frac{h}{2}, y_n + \frac{h}{2} f(x_n, y_n)\right] \quad ...(1)$$

Put,

$$n = 0$$

$$y_1 = y_0 + h\, f\left[x_0 + \frac{h}{2}, y_0 + \frac{h}{2} f(x_0, y_0)\right]$$

$$= 1 + (0.1) f\left[4 + \frac{0.1}{2}, 1 + \frac{0.1}{2} f(4,1)\right]$$

$$= 1 + 0(0.1) f((4.05.1 + (0.05)(0.05))$$

$$= 1 + (0.1) f(4.05, 1.0025)$$

$$y(4,1) = 1.0806$$

$$y_2 = y_1 + h\, f\left[x_1 + \frac{h}{2}, y_1 + \frac{h}{2} f(x_1, y_1)\right]$$

$$= 1.0806 + (0.1) f\left[4.1 + \frac{0.1}{2}, 1.0806 + \frac{0.1}{2} f(4.1.1.0806)\right]$$

$$= 1.0806 + (0.1) f(4.15, 1.1147)$$

$$y(4,2) = 1.1435$$

4. Using modified Euler's method, let us find y(0.2) and y(0.4) given y = x2 + y2, y(0) = 1 by taking h = 0.2.

Given:

$$f(x,y) = x^2 + y^2; x_0 = 0; y_0 = 1,$$

$$x_1 = 0.2;\ x_2 = 0.4;\ h = 0.2$$

To find:

$$y_1 = y(0.2)$$

Solution:

Modified Euler's formula is:

$$y_{n+1} = y_n + h\, f\left[x_n + \frac{hy}{2}, y_n + \frac{h}{2} f(x_n, y_n)\right] \quad ...(1)$$

Case (i) put n = 0 in (1)

$$y_1 = y_0 = h\, f\left[x_0 + \frac{h}{2}, y_0 + \frac{h}{2} f(x_0, y_0)\right]$$

$$= 1 + (0.2) f\left[0.1.1 + (0.1) f(0,1)\right]$$

$$= 1 + (0.2) f(0.1.1 + (0.1)\left[(0^2 + 1^2)\right]$$

$$= 1 + (0.2) f(0.1.1 + (0.1)\left[(0^2 + 1^2)\right]$$

$$= 1 + (0.2) f(0.1.1 + (0.1)\left[(0^2 + 1^2)\right]$$

$$y(0,2) = 1.244$$

To find:

$$y_2 = y(0.4)$$

Put,

n = 1 in (1),

$$y_2 = y_1 = h\, f\left[x_1 + \frac{h}{2}, y_1 + \frac{h}{2} f(x_1, y_1)\right]$$

$$= 1.244 + (0.2)\, f\left[0.3, 1.244 + (0.1)\, f\, (0.2, 1.244)\right]$$

$$= 1.244 + (0.2)\, f\left[0.3, 1.4028\right]$$

$$y(0,4) = 1.656$$

2.7 Runge-Kutta Methods

The Taylor's series method of solving differential equations numerically is restricted because of the evaluation of the higher order derivatives. Runge-kutta methods of solving initial value problems do not require the calculations of higher order derivatives and give greater accuracy.

The Runge-Kutta formula posses the advantage of requiring only the function values at some selected points. These methods agree with Taylor series solutions upto the term in h^r where r is called the order of that method.

Fourth-order Runge-Kutta Method

This is commonly used for solving initial values problem $\frac{dy}{dx} = f(x,y), y(x_0) = y_0$.

Problems

1. Using Runge-Kutta method, let us find y(0.2) if Y = $xy^2 - y^2$, y(0) = 1, y (0) - 0, h = 0.2,

Solution:

Given:

$$y = xy^2 - y^2, y(0) = 1; y(0) = 0,$$
$$x_0 = 0; y_0 = 1; = 0, x_1 = 0.2; h = 0.2,$$
$$y_1 = y(0.2)$$

Let us consider:

$$y = z;\ z_0 = 0$$

$$y = xz^2 - y^2$$

$$\frac{dy}{dx} = z = f_1(x,y,z)$$

$$\frac{dz}{dx} = xz^2 - y^2 = f_2(x,y,z)$$

By R-K Method:

$$k_1 = h\,f_1(x_0, y_0, z_0) = (0.2)\ f_1(0,1, 0) = 0$$
$$l_1 = h\,f_2(x_0, y_0, z_0) = (0.2)\ f_2(0,1, 0) = -0.2$$

$$K_2 = h\, f_1\left(x_0 + \frac{h}{2}, y_0 + \frac{k_1}{2}, z_0 + \frac{I_2}{2}\right)$$

$$= (0.2)\, f_1 (0.1,\ 1, -0.1)$$
$$= (0.2)(-\ 0.1) = -0.02$$

$$l_2 = (0.2)\, f_2 (0.1,\ 1, -0.1)$$
$$= (0.2)\ \left[(0.1)(-0.1)^2 - (1)^2\right]$$
$$l_2 = 0.1998$$

$$K_3 = h\, f_1\left(x_0 + \frac{h}{2}, y_0 + \frac{k_2}{2}, z_0 + \frac{I_2}{2}\right)$$

$$= (0.2)\, f_1\ (0.1,\ 0.99, -0.0999)$$
$$= (0.2)(-0.0999)$$
$$k_3 = -\ 0.01998$$

$$l_3 = (0.2)\, f_2\ (0.1,\ 0.99,\ -0.0999)$$
$$= (0.2)\left[(0.1)(-0.0999)^2 - (0.99)^2\right]$$
$$l_3 = -0.1958$$

$$K_4 = h\, f_1 (x_0 + h,\ y_0 + k_3,\ z_0 + l_3)$$
$$= (0.2)\, f_1\ (0{,}2,\ 0.98,\ -0.1958)$$
$$= (0.2)(-0.1958)$$
$$K_4 = -0.03916$$

$$l_4 = (0.2)\, f_2 (0.2,\ 0.98,\ -0.1958)$$
$$= (0.2)\left[(0.2)(-0.1958)^2 - (0.98)^2\right]$$
$$l_4 = -0.1905$$
$$y_0 = y_0 + \Delta y$$

$$\Delta y = \frac{1}{6}\left[k_1 + 2k_2 + 2k_3 + k_4\right]$$
$$= \frac{1}{6}\left[0 + (-0.4) + (-0.03916) - 0.03916\right]$$

$$\Delta y = -0.01985$$

$$y_1 = y(0.2) = 1 - 0.01985$$

$$y(0.2) = 0.9802$$

2. Using Runge-Kutta method of fourth Order, let us find the value of y at x = 0.2, 0.4, 0.6 given $\frac{dy}{dx} = x^3 + y,\ y(0) = 2$

Solution:

Given:

$$y' = x^3 + y = f(x, y);$$
$$x_0 = 0;\ y_0 = 2;\ x_1 = 0.2;\ x_2 = 0.4;$$
$$x_3 = 0.6;\ x_9 = 0.8;\ h = 0.2.$$

Runge Kutte formula is:

$$K_1 = h\,f(x_n, y_n)$$

$$K_2 = h\,f\left[x_n + \frac{h}{2}, y_n + \frac{k_1}{2}\right]$$

$$K_3 = h\,f\left[x_n + \frac{h}{2}, y_n + \frac{k_1}{2}\right]$$

$$K_4 = hf\left[x_n + h,\ y_n + k_3\right]$$

$$\Delta y = \frac{1}{6}\left[k_1 + 2k_2 + 2k_3 + k_4\right]$$

$$y_n + 1 = y + \Delta y$$

Case (i):

To find:

$$y_1 = y(0.2).$$

Put,

$$n = 0,$$

$$K_1 = (0.2)\ f(x_0, y_0) = (0.2)[x_0 + yo] = (0.2)[0+2]$$

$$= 0.4$$

$$K_2 = (0.2) f\left[x_0 + \frac{0.2}{2}, y_0 + \frac{0.4}{2}\right] = (0.2) f(0.1.2.2)$$

$$= (0.2)\left[(0.1)^3 + 2.2\right] = 0.4402$$

$$K_3 = (0.2) f\left[10 + \frac{0.2}{2}, 2 + \frac{0.4402}{2}\right]$$

$$= \left(0.2 f(0.1, 2.2201)\right)$$

$$= (0.2)(2.2211) = 0.44422$$

$$K_4 = (0.2)\ f\ [0.2,\ 2.44422]$$
$$= (0.2)\ [2.452221$$
$$= 0.49044$$

$$\Delta y = \frac{1}{6}\left[k_1 + 2k_2 + 2k_3 + k_4\right] = \frac{1}{6}[2.659228]$$

$$= 0.44321$$

$$y_1 = y(0,2) = y_0 + \Delta y = 2 + 0.44321$$

$$y_1 = 2.0443$$

Case (ii): To find y_2

$$K_1 = \ h\, f(x_1, y_1) = (0.2) f[0.2,\ 2.443]$$

$$= 0.4902$$

$$K_2 = h\ f\left[x_1 + \frac{h}{2}, y_1 + \frac{k_1}{2}\right] = (0.2) f(0.3, 2.6881)$$

$$= 0.5430$$

$$K_3 = hf\left[x_1 + \frac{h}{2}, y_1 + \frac{k_2}{2}\right] = (0.2)\ f(0.3, 2.7145)$$

$$=0.5483$$

$$K_4 = hf(x_1 + h, y_1 + k_3) = (0.2)f(0.4, 2.9913)$$

$$=0.6111$$

$$\Delta y = \frac{1}{6}[k_1 + 2k_2 + 2k_3 + k_4] = 0.5473$$

$$y_2 = y(0.4) = y_1 + \Delta_y = 2.443 + 0.5473 = 2.99$$

Case (iii): To find $y_3 = y(0.6)$

$$K_2 = hf\left[x_2 + \frac{h}{2}, y_2 + \frac{k_1}{2}\right] = (0.2)f(0.5, 3.2954) = 0.6841$$

$$K_3 = hf\left[x_2 + \frac{h}{2}, y_2 + \frac{k_2}{2}\right] = (0.2)f(0.5, 3.3321) = 0.6914$$

$$K_4 = hf(x_2 + h, y_2 + k_3) = (0.2)f(0.6, 3.6814) = 0.7795$$

$$\Delta y = \frac{1}{6}[k_1 + 2k_2 + 2k_3 + k_4] = 0.6902$$

$$y_3 = y(0.6) = y_2 + \Delta_y = 2.99 + 0.6902$$

$$=3.68$$

$$x_0 = 0;\ x_1 = 0.2; x_2 = 0.4; x_3 = 0.6 \ \& \ x_4 = 0.8$$

$$y_0 = 2;\ y_1 = 2.044;\ y_2 = 2.99; y_3 = 3.68$$

2.7.1 Predictor-corrector Methods

These methods required fewer evaluations of the functions to estimate the solution at a point and iteration are performed till sufficient accuracy is achieved. Estimation of error is possible and the methods are called Predictor-corrector methods.

In this type we mainly discuss Milne's and Adams-Bashforth method. In the multi-step method, to compute y_{n+1}, we need the functional values y_n, y_{n-1}, y_{n-2} and y_{n-3}.

Problems

1. Given $5xy + y^2 = 2$, y(4) = 1, y(4.1) = 1.0049, y(4.2) = 1.0097, y(4.3) = 1.0143. Let us compare y (4.4) using Milne's method.

Solution:

Given:

$$y = \frac{2-y^2}{5x}$$

$$x_0 = 4;\ x_1 = 4.1;\ x_2 = 4.2;\ x_3 = 4.3 \text{ and } x_4 = 4.4$$
$$y_0 = 1.5;\ y_1 = 1.0049;\ y_2 = 1.0097;\ y_3 = 1.0143 \text{ and } h = 0.1$$

To find: $y_4 = y(4.4)$

Milne's Predictor formula is:

$$y_{4,p} = y_0 + \frac{4h}{3}\left[2y_1' - y_2' + 2y_3'\right] \quad ...(1)$$

Now,

$$y_1' = \frac{2-y_1^2}{5x_1} = \frac{2-(1.0049)^2}{5(4.1)} = 0.0493$$

$$y_2' = \frac{2-y_2^2}{5x_2} = \frac{2-(1.0097)^2}{5(4.2)} = 0.0467$$

$$y_3' = \frac{2-y_3^2}{5x_3} = \frac{2-(1.0143)^2}{5(4.3)} = 0.0425.$$

By (1):

$$y_{4,p} = 1 + \frac{4(0.1)}{3} = \left[2(0.0493) - 0.0467 + 2(0.0452)\right]$$

$$= 1.01897$$

Now,

$$y'_4 = \frac{2-y_4^2}{5x_4} = \frac{2-(1.01897)^2}{5(4.4)} = 0.0437$$

Milne's-Corrector formula is:

$$y_{4,c} = y_2 + \frac{h}{3}\left[y'_2 + 4y'_3 + y'_4\right]$$

$$= 1.0097 + \frac{0.1}{3}\left[0.0467 + 4(0.0452) + 0.0437\right]$$

$$y_{4,c} = 1.01874$$

Hence,

$$y(4.4) = 1.01874$$

2. Using Adam's method, let us find y(2) if y′ = (x + y)/2, y(0) = 2, y(0.5) = 2.636, y(1) = 3.595, y(1.5) = 4.968.

Solution:

Given:

$$x_0 = 0;\ x_1 - 0.5;\ x_2 = 1;\ x_3 = 1.5;\ x_4 = 2;$$

$$y_0 = 2;\ y_1 = 2.636;\ y_2 = 3.595;\ y_3 = 4.968;$$

$$h = x_1 - x_0 = 0.5$$

Adam's Predictor formula is:

$$y_{n+1,p} = y_n + \frac{h}{24}[55y'_n - 59y'_{n-1} + 37y'_{n-2} - 9y'_n - 3]$$

Put,

$$n = 3$$

$$y_{4,p} = y_3 + \frac{4}{24}\left[55y'_3 - 59y'_2 + 37y'_1 - 9y'_0\right] \quad \text{...(1)}$$

$$y' = \frac{x+y}{2}$$

$$y'_0 = \frac{x_0 + y_0}{2} = \frac{0+2}{0} = 1$$

$$y_1' = \frac{x_1 + y_1}{2} = \frac{0.5 + 2.636}{2} = 1.568$$

$$y_2' = \frac{x_2 + y_2}{2} = \frac{1 + 3.595}{2} = 2.2975$$

$$y_3' = \frac{x_3 + y_3}{2} = \frac{1.5 + 4.968}{2} = 3.234$$

$$y'_{4,p} = 4.968 + \frac{(0.5)}{24}\left[55(3.234) - 59(2.2975) + 37(1.568 - 9(1))\right]$$

$$y'_{4,p} = 6.871.$$

Adam's Corrector formula is:

$$y_{n+1,c} = y_n + \frac{h}{24}\left[9y'_{n+1} + 19y'_n - 5y'_{n-1} + y'_{n-2}\right]$$

Put,

$$n = 3.$$

$$y_{4,c} = y_3 + \frac{h}{24}\left[9y'_4 + 19y'_3 - 5y'_2 + y'_1\right] \qquad \text{...(2)}$$

Now,

$$y_4' = \frac{x_4 + y_4}{2} = \frac{2 + 6.871}{2} = 4.4355$$

$$\therefore y_{4,c} = 4.968 + \left(\frac{0.5}{24}\right)\left[9(4.4355) + 19(3.234) - 5(2.2975) + 1.568\right]$$

$$= 6.873$$

$$\therefore\ y(2) = 6.873.$$

3. Using Adam's Bashforth method, let us find y(4.4) given that 5xy+y2 = 2, y(4) = 1,y(4.1) = 1.0049,y(4.2) = 1.0097 and y(4.3) = 1.0143.

Solution:

Given:

$$5xy + y^2 = 2 \Rightarrow y = \frac{2 - y^2}{5x} \qquad \text{...(1)}$$

$$x_0 = 4;\ x_1 = 4.1;\ x_2 = 4.2;\ x_3 = 4.3\ \&\ x_4 = 4.4;$$

$$y_0 = 1;\ y_1 = 1.0049;\ y_2 = 1.0097;\ y_3 = 1.0143;\ y_4 = ?$$

Here,

$$h = 0.1$$

Adam's Predictor formula is:

$$y_{n+1,p} = y_n + \frac{h}{24}\left[55y'_n - 59y'_{n-1} - 37'_{n-2} - 9y'_{n-3}\right]$$

Put,

$$n = 3$$

$$y_{4,p} = y_3 + \frac{h}{24}\left[55\ y'_3 - 59y'_2 - 37y'_1 - 9y'_0\right] \quad ...(2)$$

Now from equation (1):

$$y'_0 = \frac{2 - y_0^2}{5x_0} = 0.05$$

$$y'_1 = \frac{2 - y_1^2}{5x_1} = 0.0483$$

$$y'_2 = \frac{2 - y_2^2}{5x_2} = 0.0467$$

$$y'_3 = \frac{2 - y_3^2}{5x_3} = 0.0452$$

Substitute in (2) we get:

$$y_{4,p} = 1.0143 + \frac{0.1}{4} = \left[55(0.0452) - 59(0.0467) + 37(0.0483) - 9(0.05)\right]$$

$$y_{4,p} = 1.0186$$

Adams Corrector formula is:

$$y_{4,c} = y_3 + \frac{h}{24} = \left[9y'_4 + 19y'_3 - 5y'_2 + y'_1\right] \quad ...(3)$$

To find: y_4

Now,

$$y_4' = \frac{2 - y_4^2}{5x_4} = 0.0437$$

$$y_{4,c} = 1.0143 + \left(\frac{0.1}{24}\right) [9(0.0437) + 19(0.0452) - 5(0.0462) + 0.04831)]$$

$$= 1.0187$$

$$y(4.4) = 1.0187$$

4. Given: $\frac{dy}{dx} = xy + y^2, y(0) = 1,\ y(0,1) = 1.1169$ and y (0.2) = 1.2774, let us find.

- y(0.3) by Runge-Kutta method of forth order.
- y(0.4) by Milne's method.

Solution:

Given:

$$\frac{dy}{dx} = xy + y^2 = f(x,y);$$

$$x_0' = 0;\ x_1 = 0.1;\ x_2 = 0.2;\ x_3 = 0.4;\ x_4 = 0.4$$
$$y_0 = 1;\ y_1 = 1.1169;\ y_2 = 1.2774$$

To find:

$$y_3 = y(0.3)$$

Now,

$$k_1 = h\,f(x_2,\ y_2) = (0.1)f(0.2,\ 1.2774)$$

$$= 0.1887$$

$$k_2 = hf\left(x_2 + \frac{h}{2}, y_2 + \frac{k_1}{2}\right) = (0.1)f(0.25, 1.3718)$$

$$= 0.2225$$

$$k_3 = h\,f\left(x_2 + \frac{h}{2}, y_2 + \frac{k_2}{2}\right) = (0.1)f(0.25, 1.3887)$$

$$= 0.2274$$

$$k_4 = h\,f(x_3 + h,\ y_3 + k_3) = (0.1)\ f(0.3,\ 1.5048)$$

$$= 0.2716$$

$$y_3 = y_2 + \Delta y = y_2 + \frac{1}{6}\left[k_1 + 2k_2 + 2k_3 + k_4\right]$$

$$= 1.2774 + \frac{1}{6}[0.1887 + 210.2274 + 0.2274 + 0.2716]$$

$$= 1.5041$$

To find $y_4 = y(0.4)$ by Milne's method:

$$y_{4,p} = y_o + \frac{4h}{3}[2y_1' - y_2' + 2y_3']$$

Now,

$$y_o = x_o y_o + y^2{}_o = 1$$
$$y'_1 = x_1 y_1 + y^2{}_1 = 1.3592$$
$$y'_2 = x_2 y_2 + y^2{}_2 = 1.8872$$
$$y'_3 = x_3 y_3 + y^2{}_3 = 2.7135$$

$$\therefore y_{4,p} = 1 + \frac{4(0.1)}{3}\left[2(1.3592) - 1.8872 + (2.7135)\right]$$

$$= 1.8344$$

Now,

$$y_4 = x_4 y_4 + y^2{}_4 = 4.0988$$

By Corrector method is:

$$y_{4,C} = y_2 + \frac{h}{3}\left[y_2' + 4y_3' + y_4'\right]$$

$$= 1.2774 + \frac{0.1}{3}\left[1,8872 + 4(2,7135) + 4.0988\right]$$

$$y_4 = 1.8387$$

$$\therefore\ y(0.4) = 1.8387$$

5. Given that $\frac{dy}{dx} = 1 + y^2$; $y(0,6) = 0.6841$, y(0.4) = 0.4228, y(0.2) = 0.2027, y(0) = 0, let us find y(- 0.2) using Milne's method.

Solution:

Given:

$$y = 1 + y^2$$

	x_0	x_1	x_2	x_3	x_4
x	0.6	0.4	0.2 0	-	0.2
y	0.6841	0.4228	0.2027	0	-
	y_0	y_1	y_2	y_3	y_4

Here,

$$h = -0.2$$

Milne's Predictor formula we have:

$$y_{n+1.p} = y_{n-3} + \frac{4h}{3}\left[2y'_n - 2 - y'_{n-1} + 2y'_n\right]$$

$$y_{4,p} = y_0 + \frac{4h}{3}\left[2y'_1 - y'_2 + 2y'_3\right] \qquad ...(1)$$

$$y = 1 + y_1^2$$

$$y'_1 = 1 + y_1^2 = 1 + (0.4228)^2 = 1.1788$$

$$y_2 = 1 + y_2^2 = 1(0.2027)^2 = 1.0411$$

$$y_3 = 1 + y_3^2 = 1 + 0 = 1$$

$$\therefore y_{4,p} = 0.6841 - \frac{4(0.2)}{3}\left[2(1.1788) - 1.0411 + 2(1)\right]$$

$$= 0.6841 - 0.8844$$

$$y_{4,p} = -0.2003$$

$$y_4 = 1 + y_4^2 = 1 + (-0.2003)^2 = 1.0401$$

Milne's Corrector formula is:

$$y_{n+1,c} = y_{n-1} + \frac{h}{3}\left[y_{n-1} + 4y_n + y_{n-1}\right]$$

$$y_{4,c} = y_2 + \frac{h}{3}\left[y_2 + 4y_3 + y_4\right]$$

$$= 0.2027 - \frac{0.2}{3}\left[1.0411 + 4(1) + 1.0401\right]$$

$$= -0.2027 - 0.40451$$

$$y(-0,2) = -0.2027$$

6. Given that y + xy + y = 0, y(0) = 1, y (0) = 0 obtain y for x = 0.1, 0.2 and 0.3 by Taylor's series method and let us find the solution for y(0.4) by Milne's method.

Solution:

$$y'' = -xy' - y \quad ...(1)$$

$$x_o = 0;\ y_o = 1;$$

$$y_o = 0,\ = 0.1,\ x_2 = 0.2,\ x_3 = 0.3;\ h = 0.1$$

Put,

$$y = z \quad ...(2)$$

$$\text{in.}(1) \Rightarrow y_o = z_o = 0$$

$$z' = -xz - y \quad ...(3)$$

$y = z;\ y_o = z_o = 0$	$z = -xz - y;$ $z_o = -x_o z_o - y_o = -1$
$y'' = z = y_o = z_o = -1$	$z'' = -xz - z - y;\ z_o = 0$ $= -xz - 2z$

$y'' = z \;\; y''_0 = 0$	$z'' = -xz - z - 2z$ $= -xz - 3z;\; z''_0 = 3$
$x^v = z\,;\, y^v_0 = 3$	$2^V - -xz - z - 3z$ $= -xz - 4z;\; 2^v_0 = 0$

∴ Taylors series expansion is:

$$y(x) = y_0 + \frac{x}{1!}y'_0 + \frac{x}{2!}y''_0 + \frac{x}{3!}y'''_0 + \ldots$$

Now:

$$y(0.1) = 1 + (0.1)(0) + \frac{(0.1)^2}{2}(-1) + \frac{(0.1)^3}{6}(0) + \ldots$$

$$= 1 - \frac{(0.1)^2}{2} = 0.995$$

$$y(0.2) = 1 - \frac{(0.2)^2}{2} = 0.9802$$

$$y(0.3) = 1 - \frac{(0.3)^2}{2} = 0.956$$

$$z(x) = z_0 + \frac{x}{1!}z_0 + \frac{x^2}{2!}z_0 + \frac{x^3}{3!}z_0 + \ldots$$

$$z(0.1) = 0 + \frac{(0.1)}{1!}(-1) + 0 + \frac{(0.1)^3}{3!}(3) + \ldots$$

$$= -\frac{(0.1)}{1} + \frac{(0.1)^3}{2} = -0.0995$$

$$z(0.1) = -0.0995$$

$$z(0.2) = -\frac{(0.2)}{1} + \frac{(0.2)^3}{2} = -0.196$$

$$z(0.3) = -0.28655$$

Also from (3):

$$z'(x) = -xz - y$$

$$z'(0.1) = -(0.1)(-0.0995) - (0.995)$$

$$= -0.0985$$

$$z(0.2) = -0.941$$

$$z(0.3) = -0.87.$$

Applying Milne's Predictor formula, first to z and then to y, we obtain:

$$z[0.4] = z(0) + \frac{4}{3}(0.1)\left[2z(0.1) - z(0.2) + 2z(0.3)\right]$$

$$= 0 + \left(\frac{0.4}{3}\right)[-1.79 + 0.941 - 1.74]$$

$$= -0.3692$$

$$y(0.4) = y(0) + \frac{4}{3}(0.1)\left[2y(0.1) - y(0.2) + 2y(0.3)\right]$$

$$= 0 + \left(\frac{0.4}{3}\right)[-10.199 + 0.196 - 0.5736]$$

$$= 0.9231.\left[\because y = z; y(0.1) = z(0.1)\right]$$

Also,

$$z(0.4) = -x\left[(0.4)z(0,4) + y(0,4)\right]$$

$$= -\left[(0.4)(-0.3692) + 0.9231\right]$$

$$= -0.7754$$

Now applying Milne's corrector formula, we get:

$$z[0.4] = z(0.2) + \frac{0.1}{3}\left[z(0.2) + 4z(0.3) + z(0.4)\right]$$

$$= -0.196 + \left(\frac{0.1}{3}\right)[-0.941 - 3.48 - 0.7754]$$

$$= -0.3692$$

And,

$$y(0.4) = y(0.2) + \frac{h}{3}[y(0.2) + 4y(0.3) + y(0.4)]$$

$$= y(0.2) + \frac{h}{3}[z(0.2) + 4z(0.3) + z(0.4)]$$

$$= 0.9802 + \left(\frac{0.1}{3}\right)[-0.196 - 1.1452 - 0.3692]$$

$$= 0.9232$$

Hence,

$$y(0.4) = 0.9232$$

Formulas

1. Taylor Series method:

$$y_1 = y(x_0 + h) = y_0 + \frac{h}{1!}y'_0 + \frac{h}{2!}y''_0 + \frac{h}{3!}y'''_0 + \ldots$$

And

$$y_2 = y(x_1 + h) = y_1 + \frac{h}{1!}y'_1 + \frac{h^2}{2!}y''_1 + \frac{h^3}{3!}y'''_1 + \ldots.$$

2. Euler's method:

If,

$$\frac{dy}{dx} = f(x,y); \; y(x_0) = y_0$$

Then,

$$y = (x_0 + h) = y_0 + hf(x_0, y_0)$$

And,

$$y=(x_1+h)=y_1+hf(x_1,y_1)$$

3. Modified Euler's method:

If,

$$\frac{dy}{dx}=f(x,y);\ y(x_0)=y_0$$

Then,

$$y=(x_0+h)=y_0+h\,f\left[x_0+\frac{h}{2},y_0+\frac{h}{2}f(x_0,y_0)\right]$$

And,

$$y(x_1+h)=y_1+h\,f\left[x_1+\frac{h}{2},y_1+\frac{h}{2}f(x_1,y_1)\right]$$

4. Runge-kutta method of fourth order:

First formula:

$$k_1=hf(x_0,y_0) \qquad k_3=hf\left(x_0+\frac{h}{2},y_0+\frac{k_2}{2}\right)$$

$$k_2=hf\left(x_0+\frac{h}{2},y_0+\frac{k_1}{2}\right) \qquad k_4=hf(x_0+h,y_0+k_3)$$

And,

$$\Delta y=\frac{1}{6}\left[k_1+2k_2+2k_3+k_4\right]\ \&\ y(x_0+h)=y_0+\Delta y$$

For the second formula, replace o by 1 in the above formula.

5. Milne's Predictor formula:

$$y_{n+1}=y_{n-3}+\frac{4h}{3}\left[2y'_{n-2}-y'_{n-1}+2y'_n\right]$$

6. Milne's Corrector formula:

$$y_{n+1}=y_{n-1}+\frac{h}{3}\left[y'_{n-1}-4y'_n+2y'_{n+1}\right]$$

7. Adams-Bashforth method:

Adam's Predictor formula:

$$y_{n+1} = y_n + \frac{h}{24}\left[55y'_n - 59y'_{n-1} + 37y'_{n-2} - 9y'_{n-3}\right]$$

Adam's Corrector formula:

$$y_{n+1} = y_n + \frac{h}{24}\left[9y'_{n+1} + 19y'_n - 5y'_{n-1} + y'_{n-2}\right]$$

3

Fourier Series and Fourier Transforms

3.1 Definition of Periodic Function

Suppose f(t) is a transient function that is zero outside the interval [-T/2,T/2] or is considered to be one cycle of a periodic function.

We can obtain a sequence of coefficients by making ω a discrete variable and integrating only over the interval.

$$\hat{f}(u) = \int_{-\infty}^{\infty} f(x) e^{-j2\pi ux}\, dx = \int_{-T/2}^{T/2} f(x) e^{-j2\pi x} dx$$

$$\hat{f}_n = \hat{f}_n(n.\Delta u) = \int_{-T/2}^{T/2} f(x) e^{-jn2\pi\Delta u}\, dx \qquad \text{Where, } \Delta u = \frac{1}{T}$$

Amplitude and Spectrum of a Periodic Function

Let f be a periodic function defined for $t \in R$. If t is time variable, then we say that the periodic function f is defined in the time domain.

Further, each Fourier coefficient in the Fourier expansion of f is associated with a specific frequency $n\omega_0$. Also if the series converges to f, then the function f is completely determined by their Fourier coefficients. Therefore, we say that f is described by the Fourier coefficients in the frequency domain. The signals are generally interpreted in terms of frequencies.

For example, sound is expressed in terms of frequency as pitch whereas light is expressed in term of frequency as colour.

The sequence of Fourier coefficients C_n with $n \in Z$, which describe a function in the frequency domain, is called the spectrum of the function.

Since n assumes only integer values, the spectrum is called a discrete spectrum or line spectrum.

The sequence of absolute values of C_n, that is, $|C_n|$ is called amplitude spectrum, whereas, the sequence $\{\arg(C_n)\}$ is called phase spectrum of the function.

3.1.1 Fourier Expansion of Periodic Functions in a given Interval of Length 2

Instead of a period of 2π, many functions have an arbitrary period; say a period of 2L. In order to convert the Fourier series defined earlier for these functions, a change of variable is required:

Replace the variable x by (π/L)x: when x=L the new variable equals to π when x= -L, it equals to -π .

$$f(x)=a_o+\sum_{n=1}^{\infty}\left[a_n\cos\left(\frac{n\pi}{L}x\right)+b_n\sin\left(\frac{n\pi}{L}x\right)\right]$$

$$a_o=\frac{1}{2L}\int_{-L}^{L}f(x)dx,$$

$$a_n=\frac{1}{L}\int_{-L}^{L}f(x)\cos\left(\frac{n\pi x}{L}\right)dx, n=1,2...$$

$$b_n=\frac{1}{L}\int_{-L}^{L}f(x)\sin\left(\frac{n\pi x}{L}\right)dx, n=1,2...$$

3.1.2 Determination of Fourier Coefficients

Fourier series is an infinite series representation of periodic function in terms of the trigonometric sine and cosine functions.

Most of the single valued functions which occur in applied mathematics can be expressed in the form of Fourier series, which is in terms of sines and cosines.

Fourier series is to be expressed in terms of periodic functions- sines and cosines.

Fourier series is a very powerful method to solve ordinary and partial differential equations, particularly with periodic functions appearing as non-homogeneous terms.

We know that, Taylor's series expansion is valid only for functions which are continuous and differentiable. Fourier series is possible not only for continuous functions but also for periodic functions, functions which are discontinuous in their values and derivatives. Further, because of the periodic nature, Fourier series constructed for one period is valid for all values.

Let (x) be a function in $[0,2\pi]$. Let $f(x+2\pi)=f(x)\forall x$, then the Fourier serious of $f(x)$ is given by,

$$f(x)=\frac{a_o}{2}+\sum_{n=1}^{\infty}(a_n \cos nx+b_n \sin nx)$$

Where $a_o=\frac{1}{\pi}\int_o^{2\pi} f(x)dx$

$$a_n=\frac{1}{\pi}\int_o^{2\pi} f(x)\cos nx\,dx$$

$$b_n=\frac{1}{\pi}\int_o^{2\pi} f(x)\sin nx\,dx$$

These values a_o, a_n, b_n are called as Fourier coefficient of $f(x)$ in $[0,2\pi]$.

Let $f(x)$ be a function defined in $[-\pi,\pi]$, let $f(x+2\pi)=f(x)\forall x$ then the Founder serious of $f(x)$ is given $f(x)=\frac{a_o}{2}+\sum_{n=1}^{\infty}(a_n \cos nx+b_n \sin nx)$

Where $a_o=\frac{1}{\pi}\int_{-\pi}^{\pi} f(x)dx$

$$a_n=\frac{1}{\pi}\int_{-\pi}^{\pi} f(x)\cos nx\,dx$$

$$b_n=\frac{1}{\pi}\int_{-\pi}^{\pi} f(x)\sin nx\,dx$$

These values a_o, a_n, b_n are called as Fourier coefficients of $f(x)$ in $[-\pi,\pi]$.

Let (x) be a function in $[0,2l]$. Let $f(x+2l)=f(x)\forall x$, then the Fourier series of $f(x)$ is given by $f(x)=\frac{a_o}{2}+\sum_{n=1}^{\infty}\left(a_n \cos\frac{n\pi x}{l}+b_n \sin\frac{n\pi x}{l}\right)$

Where $a_o=\frac{1}{l}\int_o^{2l} f(x)dx$

$$a_n=\frac{1}{l}\int_o^{2l} f(x)\cos\frac{n\pi x}{l}dx$$

$$b_n=\frac{1}{l}\int_o^{2l} f(x)\sin\frac{n\pi x}{l}dx$$

These values a_o, a_n, b_n are called as Fourier coefficients of $f(x)$ in $[0,2l]$.

Let $f(x)$ be function defined in $[-l,l]$, Let $f(x+2\pi)=f(x)\forall x$ then the Fourier series of $f(x)$ is given by,

$$f(x)=\frac{a_o}{2}+\sum_{n=1}^{\infty}\left(a_n\cos\frac{n\pi x}{l}+b_n\sin\frac{n\pi x}{l}\right)$$

Where $a_o=\frac{1}{l}\int_{-l}^{l}f(x)dx$

$$a_n=\frac{1}{l}\int_{-l}^{l}f(x)\cos\frac{n\pi x}{l}dx$$

$$b_n=\frac{1}{l}\int_{-l}^{l}f(x)\sin\frac{n\pi x}{l}dx$$

These values a_o, a_n, b_n are called as Fourier coefficients of $f(x)$ in $[-l,l]$

3.2 Fourier Series of Even and Odd Functions

Fourier coefficients were calculated directly from the integral definitions:

- A function f(t) is even if f(−t) = f(t)
- A function f(t) is odd if f(−t) = −f(t)

For example, f(t) = cost is even since $f(-t)=\cos(-t)\cos t=f(t)$.

$f(t)=\sin t$ is odd since $f(-t)=\sin(-t)=-\sin t=-f(t)$.

Note that the product of two even functions f and g is even, since $f(-t)g(-t)=f(t)g(t)$.

The product of two odd functions f and g is even, since f(−t) g(−t)=(−f(t)) (−g(t)) = f(t) g(t) but the product of an odd function f with an even function g is odd, since f(−t) g(−t)=(−f(t)) g(t) = −f(t) g(t) n.

The crucial results concern the integrals of odd and even functions. From the definitions of these functions, or from their graphs, it is easily seen that:

$$\text{if } f(t) \text{ is odd then } \int_{-a}^{a}f(t)dt=0;$$
$$\text{if } f(t) \text{ is even then } \int_{-a}^{a}f(t)dt=2\int_{0}^{a}f(t)dt$$

These integral results are illustrated below:

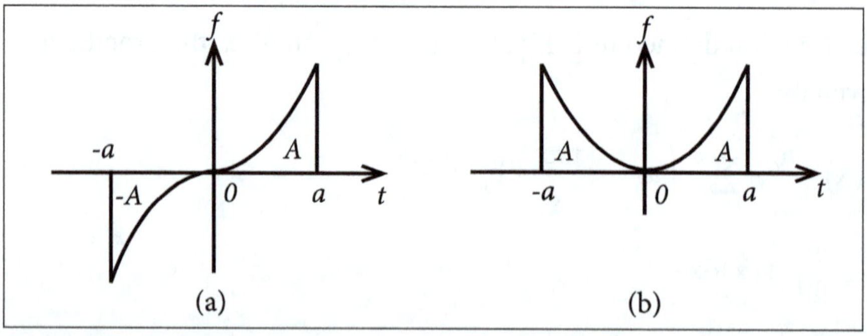

3.2.1 Fourier Series in an Arbitrary Interval

So far attention has been restricted to an interval of length of 2π. This restriction can easily be relaxed. If f(x) is periodic with a period 2L. We could also be write:

$$f(x) = \frac{a_o}{2} + \sum_{n=1}^{\infty}\left[a_n \cos\frac{n\pi x}{L} + b_n \sin\frac{n\pi x}{L}\right]$$

With:

$$a_n = \frac{1}{L}\int_{-L}^{L} f(t)\cos\frac{n\pi t}{L}dt, \quad n = 0,1,23...,$$

$$b_n = \frac{1}{L}\int_{-L}^{L} f(t)\sin\frac{n\pi t}{L}dt, \quad n = 0,1,23...,$$

3.2.2 Even and Odd Periodic Continuation

A function f(x) is even when f(x) = f(-x).

On the other hand, if f(x) = -f(-x), the function is an odd function.

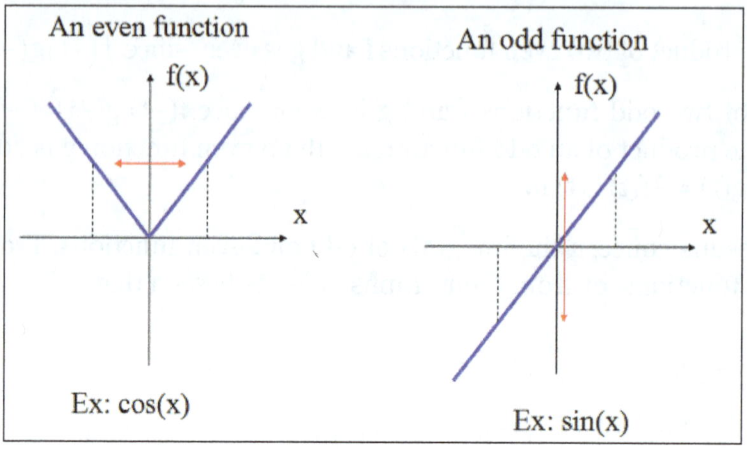

⇒ if $f(x)$ is an even function, then:

$$\int_{-L}^{L} f(x)dx = \int_{-L}^{o} f(x)dx + \int_{o}^{L} f(x)dx = \int_{L}^{o} f(-x)d(-x) + \int_{o}^{L} f(x)dx$$

$$= -\int_{L}^{o} f(x)dx = \int_{o}^{L} f(x)dx = 2\int_{o}^{L} f(x)dx$$

⇒ if $f(x)$ is an odd function, then:

$$\int_{-L}^{L} f(x)dx = \int_{-L}^{o} f(x)dx + \int_{o}^{L} f(x)dx = \int_{L}^{o} f(-x)d(-x) + \int_{o}^{L} f(x)dx$$

$$= \int_{L}^{o} f(x)dx = \int_{o}^{L} f(x)dx = o$$

⇒ if $f(x)$ is an odd function, then:

$$h(x) = f(x)g(x) \text{ is an odd function}$$

$$h(x) = f(x)g(x) = f(-x)[-g(-x)] = -[f(-x)g(-x)] = -h(-x)$$

3.2.3 Half-range Fourier Sine and Cosine Expansions

Half Range Fourier Sine Series defined in [o,Π]: The Fourier half range sine series in [o,Π] is given by:

$$f(x) = \sum_{n=1}^{\infty} b_n \sin nx,$$

Where, $b_n = \frac{2}{\pi}\int_{o}^{\pi} f(x)\sin nx\,dx$

This is Similar to the Fourier series defined for odd function in [Π,-Π]

Half Range Fourier Cosine Series defined in [o,Π]: The Fourier half range Cosine series in [o,Π] is given by:

$$f(x) = \frac{a_o}{2} + \sum_{n=1}^{\infty} a_n \cos nx$$

Where, $a_o = \frac{2}{\pi}\int_{o}^{\pi} f(x)dx$

$$a_n = \frac{2}{\pi}\int_{o}^{\pi} f(x)\cos nx\,dx$$

Half Range Fourier Sine Series defined in $[0,l]$: The Fourier half range sine series in $[0,l]$ is given by:

$$f(x)=\sum_{n=1}^{\infty} b_n \sin\frac{n\pi x}{l}$$

Where, $b_n = \frac{2}{\pi}\int_0^l f(x)\sin\frac{n\pi x}{l}dx$

This is Similar to the Fourier series defined for odd function in $[l,l]$ Half Range Fourier Cosine Series defined in $[0,l]$: The Fourier half range Cosine series in $[0,l]$ is given by:

$$f(x)=\frac{a_0}{2}=\sum_{n=1}^{\infty} a_n \cos\frac{n\pi x}{l}$$

Where

$$a_0 = \frac{2}{l}\int_0^l f(x)dx$$

$$a_n = \frac{2}{l}\int_0^l f(x)\cos\frac{n\pi x}{l}\,dx$$

This is similar to the Fourier series defined for even function in $[l,-l]$

Problems

1. Let us find the Fourier series to represent $f(x)=x^2$ in the interval $(0,2\Pi)$:

$$f(x)=\frac{a_0}{2}+\sum_{n=1}^{\infty}(a_n \cos nx + b_n \sin nx)$$

Where $a_0 = \frac{1}{\pi}\int_0^{2\pi} f(x)dx$

$$a_n = \frac{1}{\pi}\int_0^{2\pi}(x)\cos nx\,dx$$

$$b_n = \int_0^{2\pi} f(x)\sin nx\,dx$$

Here, $f(x)=x^2$

Now, $a_o = \frac{1}{\pi}\int_o^{2\pi} f(x)dx = \frac{1}{\pi}\int_o^{2\pi} x^2 dx$

$$= \frac{1}{\pi}\left[\frac{x^2}{3}\right]_o^{2\pi} = \frac{1}{3\pi}\left[(2\pi)^3 - o\right] = \frac{8}{3}\pi^2$$

$$\Rightarrow a_o = \frac{8}{3}\pi^2$$

Again, $a_n = \frac{1}{n}\int_o^{2\pi} f(x)\cos nx dx = \frac{1}{\pi}\int_o^{2\pi} \underbrace{x^2}_{u}\underbrace{\cos nx}_{v} dx$

$$= \frac{1}{\pi}\left[x^2\int \cos nx\, dx - \left\{\int \frac{d}{dx}(x)^2 (f\cos nx\, dx)dx\right\}\right]$$

$$\left[\because \int uv\,dx = u\int v\,dx - \left\{\int \frac{du}{dx}.\left(\int v\,dx\right)dx\right\}\right]$$

$$= \frac{1}{\pi}\left[x^2\left(\frac{\sin nx}{n}\right) - \left\{\int 2x\left(\frac{\sin nx}{n}\right)dx\right\}\right]_o^{2\pi}$$

$$= \frac{1}{\pi}\left[x^2\left(\frac{\sin nx}{n}\right) - \frac{2}{n}\left\{\int \underbrace{x}_{u}\underbrace{\sin nx}_{v} dx\right\}\right]_o^{2\pi}$$

$$= \frac{1}{\pi}\left[x^2\left(\frac{\sin nx}{n}\right) - \frac{2}{n}\left(-x\frac{\cos nx}{n} + \int 1.\frac{\cos nx}{n}dx\right)\right]_o^{2\pi}$$

$$= \frac{1}{\pi}\left[x^2\left(\frac{\sin nx}{n}\right) - \frac{2}{n}\left(-x\frac{\cos nx}{n} + \frac{1}{n}\int \cos nx\, dx\right)\right]_o^{2\pi}$$

$$= \frac{1}{\pi}\left[x^2\left(\frac{\sin nx}{n}\right) - \frac{2}{n}\left(-x\frac{\cos nx}{n} + \frac{1}{n}\frac{\sin nx}{n}\right)\right]_o^{2\pi}$$

$$= \frac{1}{\pi}\left[x^2\left(\frac{\sin nx}{n}\right) + \frac{2}{n^2}x\cos nx - \frac{2}{n^2}\sin nx\right]_o^{2\pi}$$

$$= \frac{4}{n^2}\left[\begin{matrix}\because \cos 2n\pi = 1\\ \sin 2n\pi = o\end{matrix}\right]$$

$$\Rightarrow a_n = \frac{4}{n^2}$$

Again, $b_n = \frac{1}{\pi}\int_o^{2\pi} f(x)\sin nx dx = \frac{1}{\pi}\int_o^{2\pi} \underbrace{x^2}_{u} \underbrace{\sin nx}_{v} dx$

$$= \frac{1}{\pi}\left[x^2\int \sin nx\, dx - \left\{\int \frac{d}{dx}(x^2)\left(\int \sin nx\, dx\right)dx\right\}\right]$$

$$\left[\because \int uv\, dx = u\int v\, dx - \left\{\int \frac{du}{dx}.\left(\int v\, dx\right)dx\right\}\right]$$

$$= \frac{1}{\pi}\left[x^2\left(\frac{-\cos nx}{n}\right) - \left\{\int 2x\left(-\frac{\cos nx}{n}\right)dx\right\}\right]_o^{2\pi}$$

$$= \frac{1}{\pi}\left[x^2\left(\frac{-\cos nx}{n}\right) + \frac{2}{n}\left\{\int \underbrace{x}_{u}\underbrace{\cos nx}_{v}\, dx\right\}\right]_o^{2\pi}$$

$$= \frac{1}{\pi}\left[-x^2\left(\frac{\cos nx}{n}\right) + \frac{2}{n}\left\{x\frac{\sin nx}{n} + \int 1.\frac{\sin nx}{n}dx\right\}\right]_o^{2\pi}$$

$$= \frac{1}{\pi}\left[-x^2\left(\frac{\cos nx}{n}\right) + \frac{2}{n}\left\{x\frac{\sin nx}{n} + \frac{1}{n}\int \sin nx\, dx\right\}\right]_o^{2\pi}$$

$$= \frac{1}{\pi}\left[-x^2\left(\frac{\cos nx}{n}\right) + \frac{2}{n}\left(x\frac{\sin nx}{n} + \frac{1}{n}\frac{\cos nx}{n}\right)\right]_o^{2\pi}$$

$$= \frac{1}{\pi}\left[-x^2\left(\frac{\cos nx}{n}\right) + \frac{2}{n^2}x\sin nx + \frac{2}{n^2}\cos nx\right]_o^{2\pi}$$

$$= -\frac{4\pi}{n}\left[\begin{matrix}\because \cos 2n\pi = 1 \\ \sin 2n\pi = o\end{matrix}\right]$$

$$\Rightarrow b_n = \frac{4\pi}{n}$$

$$f(x) = \frac{a_o}{2} + \sum_{n=1}^{\infty}(a_n \cos nx + b_n \sin nx)$$

$$\therefore f(x) = x^2 = \frac{\frac{8\pi^2}{3}}{2} + \sum_{n=1}^{\infty}\left(\frac{4}{n^2}\cos nx - \frac{4\pi}{n}\sin nx\right)$$

$$\Rightarrow x^2 = \frac{4\pi^2}{3} + \sum_{n=1}^{\infty}\left(\frac{4}{n^2}\cos nx - \frac{4\pi}{n}\sin nx\right)$$

This is the Fourier series for the function $f(x) = x^2$

2. Let us find the Fourier series of the periodic function defined as $f(x) = \begin{cases} -\pi & ; -\pi < x < 0 \\ x & ; 0 < x < \pi \end{cases}$

Hence we shall deduce that $\frac{1}{1^2} + \frac{1}{3^2} + \frac{1}{5^2} + \ldots = \frac{\pi^2}{8}$.

$$f(x) = \frac{a_o}{2} + \sum_{n=1}^{\infty}(a_n \cos nx + b_n \sin nx)$$

Where, $a_o = \frac{1}{\pi}\int_{-\pi}^{\pi} f(x)\,dx$

$$a_n = \frac{1}{\pi}\int_{-\pi}^{\pi} f(x)\cos nx\,dx$$

$$b_n = \frac{1}{\pi}\int_{-\pi}^{\pi} f(x)\sin nx\,dx$$

Here, $f(x) = \begin{cases} -\pi; -\pi < 0 < 0 \\ x; 0 < x < \pi \end{cases}$

Now, $a_o = \frac{1}{\pi}\int_{-\pi}^{\pi} f(x)\,dx = \frac{1}{\pi}\left[\int_{-\pi}^{o} f(x)\,dx + \int_{o}^{\pi} f(x)\,dx\right]$

$$= \frac{1}{\pi}\left[\int_{-\pi}^{o}(-\pi)\,dx + \int_{o}^{\pi} x\,dx\right]$$

$$= \frac{1}{\pi}\left[(-\pi)\int_{-\pi}^{o} dx + \int_{o}^{\pi} x\,dx\right]$$

$$= \frac{1}{\pi}\left[(-\pi)[x]_{-\pi}^{o} + \left[\frac{x^2}{2}\right]_{o}^{\pi}\right] = \frac{1}{\pi}\left[(-\pi)(\pi) + \frac{\pi^2}{2}\right]$$

$$= \frac{1}{\pi}\left[-\pi^2 + \frac{\pi^2}{2}\right] = -\frac{\pi}{2}$$

$$a_o = -\frac{\pi}{2}$$

Also, $a_n = \frac{1}{\pi}\int_{-\pi}^{\pi} f(x)\cos nx\,dx$

$$= \frac{1}{\pi}\left[\int_{-\pi}^{o} f(x)\cos nx\,dx + \int_{o}^{\pi} f(x)\cos nx\,dx\right]$$

$$= \frac{1}{\pi}\left[\int_{-\pi}^{o} (-\pi)\cos nx\,dx + \int_{o}^{\pi} x\cos nx\,dx\right]$$

$$= -\left[-\pi\int_{p}(\cos nx)dx + \int_{o} x\cos nx dx\right]$$

$$= \frac{1}{\pi}\left[-\pi\left(\frac{\sin nx}{n}\right)_{-\pi}^{o} + \left\{x\left(\frac{\sin nx}{n}\right) - \int 1\left(\frac{\sin nx}{n}dx\right)\right\}_{o}^{\pi}\right]$$

$$= \frac{1}{\pi}\left[-\frac{\pi}{n}(\sin nx)_{-\pi}^{o} + \left\{\frac{x\sin nx}{n} - \frac{1}{n}\int \sin nx\,dx\right\}_{o}^{\pi}\right]$$

$$= \frac{1}{\pi}\left[-\frac{\pi}{n}(\sin nx)_{-\pi}^{o} + \left\{\frac{x\sin nx}{n} - \frac{1}{n}\frac{(-\cos nx)}{n}\right\}_{o}^{\pi}\right]$$

$$= \frac{1}{\pi}\left[-\frac{\pi}{n}(\sin nx)_{-\pi}^{o} + \left\{\frac{x\sin nx}{n} + \frac{1}{n^2}\cos nx\right\}_{o}^{\pi}\right]$$

$$= \frac{1}{\pi}\left[-\frac{\pi}{n}\left[o - \sin(-n\pi)\right] + \left\{\left(\frac{\pi\sin n\pi}{n} + \frac{1}{n^2}\cos n\pi\right) - \left(o + \frac{1}{n^2}\cos no\right)\right\}\right]$$

$$= \frac{1}{\pi}\left[-\frac{\pi}{n}\sin n\pi + \left\{\left(\frac{\pi\sin n\pi}{n} + \frac{1}{n^2}\cos n\pi\right) - \frac{1}{n^2}.1\right\}\right] \quad \left[\because \begin{matrix}\cos(-\theta) = \cos\theta \\ \sin(-\theta) = \sin\theta\end{matrix}\right]$$

$$\Rightarrow a_n = \frac{1}{\pi}\left[-\frac{\pi}{n}(o) + \left\{\left(\frac{\pi(o)}{n} + \frac{1}{n^2}(-1)^n\right) - \frac{1}{n^2}\right\}\right] \quad \left[\because \begin{matrix}\sin n\pi = o \\ \& \\ \cos n\pi = (-1)^n\end{matrix}\right]$$

$$= \frac{1}{\pi}\left[\frac{(-1)^n}{n^2} + \frac{1}{n^2}\right] = \frac{1}{\pi n^2}\left[(-1)^n - 1\right]$$

$$\Rightarrow a_n = \frac{1}{\pi n^2}\left[(-1)^n - 1\right]$$

Again, $b_n = \frac{1}{\pi}\int_{-\pi}^{\pi} f(x)\sin nx\,dx$

$$= \frac{1}{\pi}\left[\int_{-\pi}^{0} f(x)\sin nx\,dx + \int_{0}^{\pi} f(x)\sin nx\,dx\right]$$

$$= \frac{1}{\pi}\left[\int_{-\pi}^{0} (-\pi)\sin nx\,dx + \int_{0}^{\pi} x\sin nx\,dx\right]$$

$$= -\left[-\pi\int_{-\pi}(\sin nx)\,dx + \int\ x\sin nx\,dx\right]$$

$$= \frac{1}{\pi}\left[-\pi\left(-\frac{\cos nx}{n}\right)_{-\pi}^{0} + \left\{x\left(-\frac{\cos nx}{n}\right) - \int 1\left(-\frac{\cos nx}{n}\right)dx\right\}_{0}^{\pi}\right]$$

$$= \frac{1}{\pi}\left[\frac{\pi}{n}(\cos nx)_{-\pi}^{0} + \left\{-\frac{x\cos nx}{n} + \frac{1}{n}\int \cos nx\,dx\right\}_{0}^{\pi}\right]$$

$$= \frac{1}{\pi}\left[\frac{\pi}{n}(\cos nx)_{-\pi}^{0} + \left\{-\frac{x\cos nx}{n} + \frac{1}{n}\frac{(\sin nx)}{n}\right\}_{0}^{\pi}\right]$$

$$= \frac{1}{\pi}\left[\frac{\pi}{n}(\cos nx)_{-\pi}^{0} + \left\{-\frac{x\cos nx}{n} + \frac{1}{n^2}\sin nx\right\}_{0}^{\pi}\right]$$

$$= \frac{1}{\pi}\left[\frac{\pi}{n}\left[1-\cos(-n\pi)\right] + \left\{\left(-\frac{\pi\cos n\pi}{n} + \frac{1}{n^2}\sin n\pi\right) - \left(0 + \frac{1}{n^2} + \sin n0\right)\right\}\right]$$

$$= \frac{1}{\pi}\left[\frac{\pi}{n}\left[1-\cos n\pi\right] - \frac{\pi\cos n\pi}{n}\right] = \frac{1}{n}(1-2\cos n\pi)$$

$$\left[\because \begin{array}{l}\cos(-\theta)=\cos\theta \\ \sin(-\theta)=\sin\theta\end{array} \quad \& \quad \begin{array}{c}\sin n\pi = 0 \\ \& \\ \cos n\pi = (-1)^n\end{array}\right]$$

$$\Rightarrow b_n = \frac{1}{n}(1-2\cos n\pi)$$

Hence the Fourier series for given f(x) is given by:

$$f(x)=\frac{\frac{-\pi}{2}}{2}\sum_{n=1}^{\infty}\left(\frac{1}{\pi n^2}\left[(-1)^n-1\right]\cos nx+\frac{1}{n}(1-2\cos n\pi)\sin nx\right)$$

$$\Rightarrow f(x)=-\frac{\pi}{4}+\sum_{n=1}^{\infty}\left(\frac{1}{\pi n^2}\left[(-1)^n-1\right]\cos nx+\frac{1}{n}(1-2\cos n\pi)\sin nx\right)$$

$$\Rightarrow f(x)=-\frac{\pi}{4}-\frac{2}{\pi}\left(\cos x+\frac{\cos 3x}{3^2}+\frac{\cos 5x}{5^2}+...\right)+\left(3\sin x-\frac{\sin 2x}{2}+\frac{\sin 3x}{3}-\frac{\sin 4x}{4}+...\right)$$

Put x=0 in the above equation f(x) we get:

$$f(0)=\frac{\pi}{4}-\frac{2}{\pi}\left(1+\frac{1}{3^2}+\frac{1}{5^2}+...\right)$$

Since, $f(x)$ is discontinuous at $x=0$, $\begin{cases} f(0-0)=-\pi \\ f(0+0)=0 \end{cases}$

$$\Rightarrow f(0)=\frac{1}{2}\left[f(0-0)+f(0+0)\right]$$

$$\Rightarrow f(0)=\frac{1}{2}(-\pi)=-\frac{\pi}{2}$$

Hence, $f(0)=-\frac{\pi}{2}=-\frac{\pi}{4}-\frac{2}{\pi}\left(1+\frac{1}{3^2}+\frac{1}{5^2}+...\right)$

$$\Rightarrow \frac{1}{1^2}+\frac{1}{3^2}+\frac{1}{5^2}+...=\frac{\pi^2}{8}$$

Hence the result.

3. Let us expand the function $f(x)=x^2$ as Fourier series in [-Π, Π].Hence we shall deduce that $\frac{1}{1^2}+\frac{1}{2^2}+\frac{1}{3^2}+\frac{1}{4^2}+...=\frac{\pi^2}{6}$

Solution:

$$f(x)=\frac{a_0}{2}+\sum_{n=1}^{\infty}(a_n\cos nx+b_n\sin nx)$$

Where, $a_o = \frac{1}{\pi}\int_{-\pi}^{\pi} f(x)dx$

$$a_n = \frac{1}{\pi}\int_{-\pi}^{\pi} f(x)\cos nx\,dx$$

$$b_n = \frac{1}{\pi}\int_{-\pi}^{\pi} f(x)\sin nx\,dx$$

Here,

$$f(x) = x^2$$

Now, $a_o = \frac{1}{\pi}\int_{-\pi}^{\pi} f(x)dx$

$$= \frac{1}{\pi}\int_{-\pi}^{\pi} x^2dx = \frac{2}{\pi}\int_{o}^{\pi} x^2dx$$

$$= \frac{2}{\pi}\left[\frac{x^2}{3}\right]_o^{\pi} = \frac{2\pi^2}{3}$$

$$\Rightarrow a_o = \frac{2\pi^2}{3}$$

Again, $a_n = \frac{1}{\pi}\int_{-\pi}^{\pi} f(x)\cos nx\,dx$

$$= \frac{1}{\pi}\int_{-\pi}^{\pi} x^2 \cos nx\,dx$$

$$= \frac{2}{\pi}\int_{o}^{\pi} x^2 \cos nx\,dx \quad \left[\because f(x) \text{ is even} \Rightarrow \int_{-a}^{a} f(x)dx = 2\int_{o}^{a} f(x)dx\right]$$

$$= \frac{2}{\pi}\left[\frac{x^2 \sin nx}{n} + \frac{2x\cos nx}{n^2} - \frac{2x\sin nx}{n^3}\right]_o^{\pi} = \frac{4}{n^2}(-1)^n$$

$$\Rightarrow a_n = \frac{4}{n^2}(-1)^n$$

Again, $a_n = \frac{1}{\pi}\int_{-\pi}^{\pi} f(x)\sin nx\,dx$

$$= \frac{1}{\pi}\int_{-\pi}^{\pi} x^2 \sin nx\,dx$$

$$= 0 \quad \left[\because f(x) \text{ is odd} \Rightarrow \int_{-a}^{a} f(x)dx = 0\right]$$

Hence, the Fourier series for given $f(x)$ is given by:

$$f(x) = x^2 = \frac{\left(\frac{2\pi}{3}\right)^2}{2} + \sum_{n=1}^{\infty} \frac{4}{n^2}(-1)^n \cos nx$$

$$\Rightarrow x^2 = \frac{\pi^2}{3} + 4\left(-\cos x + \frac{\cos 2x}{2^2} - \frac{\cos 3x}{3^2} + \frac{\cos 4x}{4^2} - \ldots\right)$$

Put $x = \pi$ in the above equation, we get:

$$\Rightarrow \pi^2 = \frac{\pi^2}{3} + 4\left(-\cos \pi + \frac{\cos 2\pi}{2^2} - \frac{\cos 3\pi}{3^2} + \frac{\cos 4\pi}{4^2} - \ldots\right)$$

$$\Rightarrow \pi^2 - \frac{\pi^2}{3} = 4\left(1 + \frac{1}{2^2} + \frac{1}{3^2} + \frac{1}{4^2} + \ldots\right)$$

$$\Rightarrow \frac{2\pi^2}{3} = 4\left(1 + \frac{1}{2^2} + \frac{1}{3^2} + \frac{1}{4^2} + \ldots\right)$$

$$\Rightarrow \frac{\pi^2}{6} = 1 + \frac{1}{2^2} + \frac{1}{3^2} + \frac{1}{4^2} + \ldots$$

Hence the result.

3.3 Fourier Integral Theorem

If f(x) satisfies the Dirichlet's conditions for expansion of Fourier series in (-c, c) and $\int_{-\infty}^{\infty} |f(x)| dx$ converges, then

$$f(x) = \frac{1}{\pi}\int_{0}^{\infty}\int_{-\infty}^{\infty} f(t).\cos\lambda(t-x)dt\,d\lambda$$

(Which is known as Fourier integral of f(x)).

Proof: Since f(x) satisfies Dirichlet's conditions in (—c, c), we have the Fourier series for f(x) as:

$$f(x)=\frac{1}{2}a_o+\sum_{n=1}^{\infty}\left(a_n\cos\frac{n\pi x}{c}+b_n\sin\frac{n\pi x}{c}\right) \quad \text{...(i)}$$

$$a_o=\frac{1}{c}\int_{-c}^{c}.f(t)dt$$

$$a_n=\frac{1}{c}\int_{-c}^{c}.f(t)\cos\frac{n\pi t}{c}dt$$

$$b_n=\frac{1}{c}\int_{-c}^{c}.f(t)\sin\frac{n\pi t}{c}dt$$

Substituting the values of a_n, b_n, a_o in above we get:

$$f(x)=\frac{1}{2c}\int_{-c}^{c}f(t)dt+\frac{1}{c}\sum\int_{-c}^{c}f(t)\cos\frac{n\pi(t-x)}{c}.dt \quad \text{...(ii)}$$

Since $\int_{-\infty}^{\infty}|f(x)|\,dx$ converges i.e, $f(x)$ is absolutely on x-axis the first term on the right hand side $\to o \text{ as } c\to\infty$,since:

$$\left|\frac{1}{2c}\int_{-c}^{c}f(t)dt\right|\le\frac{1}{2c}\int_{-\infty}^{\infty}|f(t)|\,dt$$

And the second term on the right hand side approaches follow:

$$\underset{c\to\infty}{\mathrm{Lt}}\frac{1}{c}\sum_{n=1}^{\infty}\int_{-\infty}^{\infty}f(t)\cos\frac{n\pi(t-x)}{c}dt$$

$$=\underset{x/c\to o}{\mathrm{Lt}}\frac{1}{\pi}\sum_{n=1}^{\infty}\int_{-\infty}^{\infty}f(t)\cos\frac{n\pi(t-x)}{c}dt$$

Let $\pi/c=\Delta\lambda$

$$\therefore\ f(x)=\underset{\Delta\lambda\to o}{\mathrm{Lt}}\frac{1}{\pi}\sum_{n=1}^{\infty}f(t).\cos n(t-x)\Delta\lambda\,dt \quad \text{...(iii)}$$

Which resembles the Riemann sum of definite integral.

$\therefore$ as $c\to\infty$, (iii) becomes:

$$f(x)=\frac{1}{\pi}\int_{o}^{\infty}\int_{-\infty}^{\infty}f(t).\cos\lambda(t-x)dt.d\lambda$$

The Fourier integral representation of f(x) converges to f(x) at a point or continuity and to $\left[f(x+0)+f(x-0)\right]/2$ at a point of discontinuity.

3.3.1 Fourier Sine and Cosine Integrals

If the function $f(x)$ is even

$$\Rightarrow A(w)=\frac{1}{\pi}\int_{-\infty}^{\infty} f(v)\cos(wv)dv$$

$$=\frac{1}{\pi}\int_{-\infty}^{0}\left(dv+\frac{1}{\pi}\right)\int_{0}^{\infty}\left(dv+\frac{2}{\pi}\right)\int_{0}^{\infty} f(v)\cos(wv)dv$$

$$B(w)=0$$

$$f(x)=\int_{0}^{\infty} A(w)\cos(wx)dw\text{: Fourier cosine Integral}$$

If the function $f(x)$ is odd

$$\Rightarrow B(w)=\frac{1}{\pi}\int_{-\infty}^{\infty} f(v)\sin(wv)dv$$

$$A(w)=0,$$

$$f(x)=\int_{0}^{\infty} B(w)\sin(wx)dw\text{ : Fourier sine Integral.}$$

Problems

1. Let us find the Fourier sine and cosine integral of $f(x)=e^{-kx}, x>0, k>0$.

Solution:

Clearly f(x) is differentiable and is absolutely integrable over $(0,\infty)$ To obtain Fourier cosine representation, we have:

$$A(w)=\frac{2}{\pi}\int_{0}^{\infty} f(u)\cos wu\,du=\frac{2}{\pi}\int_{0}^{\infty} e^{-ku}\cos wu\,du$$

$$I=\int_{0}^{\infty} e^{-ku}\cos wu\,du=\left[\frac{e^{-ku}}{w^2+k^2}(w\sin wu-k\cos wu)\right]_{0}^{\infty}$$

When u tends to infinity, it becomes zero, and when u tends to zero, it becomes $-k/\left(w^2+k^2\right)$, since k > 0. Thus $I=k/\left(w^2+k^2\right)$ and hence:

$$A(w)=\frac{2}{\pi}\int_0^{\infty} e^{-ku}\cos wu\,du=\frac{2k}{\pi\left(w^2+k^2\right)}$$

Thus the Fourier cosine integral $f(x)=\int_0^{\infty} A(w)\cos wx\,dx$ becomes:

$$e^{-kx}=\frac{2k}{\pi}\int_{\pi}^{\infty}\frac{\cos wx}{k^2+w^2}dx, k>0$$

On the similar lines, to obtain the Fourier sine integral representation of f(x), we have:

$$B(w)=\frac{2}{\pi}\int_0^{\infty} e^{-ku}\sin wu\,du=\frac{2w}{\pi\left(k^2+w^2\right)}$$

and thus:

$$f(x)=e^{-kx}=\frac{2}{\pi}\int_0^{\infty}\frac{w\sin wx}{k^2+w^2}dw, w>0$$

2. Let $f(x)=\begin{cases}1, & 0<x<1\\ 0, & x\geq 1\end{cases}$ Show that $\int_0^{\infty}\frac{\sin t}{t}dt=\pi/2$ by using the Fourier cosine integral representation of f.

Solution:

The function f(x) is piecewise smooth and is also absolutely integrable over (0, ∞). To obtain Fourier cosine representation, we have:

$$A(w)=\frac{2}{\pi}\int_0^{\infty} f(u)\cos wu\,du=\frac{2}{\pi}\int_0^{1}\cos wu\,du=\left[\frac{\sin wu}{w}\right]^1=\frac{2\sin w}{\pi w}$$

Thus the Fourier cosine integral representation of $f(x)$ is:

$$f(x)=\int_0^{\infty} A(w)\cos wx\,dw=\frac{2}{\pi}\int_0^{\infty}\frac{\sin w}{w}\cos wx\,dw$$

The representation converges to $f(x)$ for every x in $(0,\infty)$ except at the point x=1 which is a point of discontinuity of $f(x)$.

At x=1, the representation coverage to:

$$\frac{f(x+0)+f(x-0)}{2}=\frac{f(1+0)+f(1-0)}{2}=\frac{1}{2}$$

Therefore $\int_0^\infty \frac{\sin w}{w}\cos wx\,dw = \begin{Bmatrix} \pi/2, & 0<x<1 \\ \pi/4, & x=1 \\ 0. & x>1 \end{Bmatrix}$

For x=1 gives

$$\int_0^\infty \frac{\sin w\cos w}{w}dw = \frac{\pi}{4}, \text{ or } \int_0^\infty \frac{\sin 2w}{2w}dw = \frac{\pi}{4}$$

Setting 2w = t in it, obtain $\int_0^\infty \frac{\sin t}{t}dt = \frac{\pi}{2}$.

3. Let's solve the integral equation $\int_0^\infty f(x)\sin ax\,dx = e^{-a}$, where is constant.

Solution:

The given integral is Fourier sine integral representation.

Let $\quad f(x) = \int_0^\infty A(w)\sin wx\,dw,$

Where $A(w) = \frac{2}{\pi}\int_n^\infty f(u)\sin wu\,du$

Comparing, we get:

$$w = a \text{ and } \frac{\pi A(w)}{2} = e^{-a}, \text{ thus } A(w) = \frac{2}{\pi}e^{-w},$$

$$f(x) = \frac{2}{x}\int_0^\infty e^{-w}\sin wx\,dw$$

Let us consider:

$$I = \int_0^\infty e^{-w}\sin wx\,dw = \left[-\frac{e^{-w}}{1+x^2}(x\cos wx + \sin wx)\right]_0^\infty$$

When ω tends to infinity, this becomes zero and at $\omega = 0$, it is $-x/(1+x^2)$ and thus $I = x/(1+x^2)$. Using the above equation, we get:

$$f(x) = \frac{2x}{\pi(1+x^2)}, \quad x>0$$

3.4 Fourier Transforms

Problems

1. Let us solve transform pair of f(x) = 1 – | x | if | x | < 1 = 0 if | X | 1 and hence find the value of $\int_0^\infty \left(\frac{\sin t}{t}\right) dt$ and $\int_0^\infty \left(\frac{\sin t}{t}\right)^4 dt$.

Solution:

$$F(x) = \frac{1}{\sqrt{2x}} \int_{-1}^{+1} e^{isx(x)dx}$$

$$= \frac{1}{\sqrt{2\pi}} \int_{-1}^{+1} (1-|x|) e^{isx} dx \qquad = \frac{1}{\sqrt{2\pi}} \int_{-1}^{+1} (1-|x|)(\cos sx + i \sin x) dx$$

$$= \frac{1}{\sqrt{2\pi}} \int_{-1}^{+1} (1-|x|)(\cos sx\, dx)$$

$$= \frac{2}{\sqrt{2\pi}} \int_0^1 (1-|x|) \cos sx\, dx$$

$$= \frac{2}{\sqrt{2}} \int_0^1 (1-|x|) \cos sx\, dx$$

$$= \sqrt{2\pi} \left[(1-x)\left(\frac{\sin sx}{s}\right) - (-1)\left(-\frac{\cos sx}{s^2}\right) \right]_0^1$$

$$= \sqrt{\frac{2}{\pi}} \left[\frac{1-\cos s}{s^2} \right]$$

By sine inverse Transform is:

$$f(x) = \frac{1}{\sqrt{2\pi}} \int_{-1}^{+1} \sqrt{\frac{2}{\pi}} \left(\frac{1-\cos s}{s^2}\right) e^{-isx} ds$$

$$= \frac{2}{\pi} \int_0^\infty \left(\frac{1-\cos}{s^2}\right) \cos sx\, dx$$

Put x = 0 we have:

$$f(0) = \frac{2}{\pi} \int_0^\infty \frac{1-\cos s}{s^2} ds \Rightarrow \int_0^\infty \left(\frac{\sin t}{t}\right)^2 dt = \pi/2$$

Using Parseval's identities:

$$\int_{-1}^{+1}\left(Ff(x)\right)^2 dx=\left(\int_{\infty}^{-\infty}f(x)\right)^2 dx$$

$$\Rightarrow \frac{2}{\pi}\int_{-1}^{+1}\left(\frac{1-\cos s^2}{s}\right)ds=\int_{-1}^{+1}(1-|x|)^2 dx$$

$$=2\int_{0}^{1}(1-|x|)^2 dx$$

$$\therefore \int_{0}^{\infty}\left(\frac{\sin t}{t}\right)^4 dt=\pi/3$$

2. Let us solve an example of Fourier transform e $\frac{-x^2}{2}$ is e $\frac{-s^2}{2}$.

Solution:

$$F[f(x)]=\frac{1}{\sqrt{2\pi}}\int_{-\infty}^{\infty}f(x)e^{isx}dx$$

$$F\left[e-\frac{x^2}{2}\right]=\frac{1}{\sqrt{2\pi}}\int_{-\infty}^{\infty}e^{-\frac{x^2}{2}}e^{isx}dx$$

$$=\frac{1}{\sqrt{2\pi}}\int_{-\infty}^{\infty}e^{-\left(\frac{x^2}{2}-isx\right)dx} \quad ...(1)$$

Consider $\frac{x^2}{2}-isx=\frac{1}{2}\left(x^2-2isx\right)$

$$=\frac{1}{2}\left[x^2-2isx+(is)^2-(is)^2\right]$$

$$=\frac{1}{2}\left[(x-is)^2+s^2\right]$$

Substituting in (1):

$$F\left[e-\frac{x^2}{a}\right]=\frac{1}{\sqrt{2\pi}},\int_{-\infty}^{\infty}e^{-\frac{1}{2}(x-is)^2}.e^{-\frac{S^2}{2}}dx$$

$$=\frac{e^{\frac{s^2}{2}}}{\sqrt{2\pi}}\int_{-\infty}^{\infty}e^{-\frac{1}{2}}(x-is)^2 dx$$

Put $t=\frac{x-is}{\sqrt{2}};dt=\frac{dx}{\sqrt{2}}$

$x\rightarrow-\infty;\ t\rightarrow-\infty$

$x\rightarrow\infty;t\rightarrow\infty$

$$F\left[e^{-\frac{x^2}{2}}\right]=\frac{e^{-s^2/2}}{\sqrt{2\pi}}\int_{-\infty}^{\infty}e^{-t^2}\sqrt{2}\,dt$$

$$=\frac{e^{-s^2/2}}{\sqrt{\pi}}\int_{-\infty}^{\infty}e^{-t^2}dt$$

$$F\left[e^{-\frac{x^2}{2}}\right]=e^{-s^2/2}\qquad\left(\because\int_{-\infty}^{\infty}e^{-t^2}dt=\sqrt{\pi}\right)$$

3. Let us see the Fourier transform of e $^{a|x|}$, a> 0. Hence deduce that

$$F\left(xe^{-1|x|}\right)=i\sqrt{\frac{2}{\pi}}\frac{2as}{\left(a^2+s^2\right)^2}$$

Solution:

$$F\left[f(x)\right]=\frac{1}{\sqrt{2\pi}}\int_{-\infty}^{\infty}f(x)e^{isx}dx$$

$$=\frac{1}{\sqrt{2\pi}}\int_{-\infty}^{x}e^{-a|x|}e^{isx}dx$$

$$=\frac{2}{\sqrt{2\pi}}\int_{0}^{\infty}e^{-ax}\cos sx\,dx$$

$$=\frac{2}{\sqrt{2\pi}}\frac{a}{a^2+s^2}$$

$$F\left[xf(x)\right]=-i\frac{d}{ds}F(s)$$

$$=-i\frac{d}{ds}\frac{2}{\sqrt{2\pi}}\frac{a}{a^2+s^2}$$

$$= i\sqrt{\frac{2}{\pi}}\frac{2as}{\left(a^2+s^2\right)}$$

4. Let us solve the Fourier transform equation.

$$f(x)=\begin{cases} a^2-x^2, \text{if} \left|x\right|<a \\ o, \quad \text{if} \left|x\right|>a>o \end{cases}$$

$$\int_o^\infty \frac{\sin t - t\cos t}{t^3}dt=\frac{\pi}{4}$$

Solution:

We know that:

$$F\left[f(x)\right]=\frac{1}{\sqrt{2\pi}}\int_{-\infty}^{\infty} e^{isx}dx$$

$$F\left[f(x)\right]=\frac{1}{\sqrt{2\pi}}\int_{-a}^{a}\left(a^2-x^2\right)\left(\cos sx+\sin sx\right)dx$$

$$=\frac{1}{\sqrt{2\pi}}\int_{-a}^{a}\left(a^2-x^2\right)\cos sx\, dx+\frac{i}{\sqrt{2\pi}}\int_{-a}^{a}\left(a^2-x^2\right)\sin sx\, dx$$

$$=\frac{1}{\sqrt{2\pi}}\int_{o}^{a}\left(a^2-x^2\right)\cos sx\, dx+i(o)$$

$\because \left(a^2-x^2\right)$ sin sx is odd function.

$$=\sqrt{\frac{2}{x}}\left[\left(a^2-x^2\right)\left(\frac{\sin sx}{s}\right)-(2x)\left(-\frac{\cos sx}{s^2}\right)+(-2)\left(-\frac{\sin sx}{s^3}\right)\right]_o^a$$

$$F\left[f(x)\right]=2\sqrt{\frac{2}{\pi}}\left(\frac{\sin as - as\cos as}{s^3}\right)$$

By inverse Fourier transform:

$$f(x)=\frac{1}{\sqrt{2\pi}}\int_{-\infty}^{\infty} F\left[f(x)\right]e^{-isx}dz$$

$$=\frac{1}{\sqrt{2\pi}}\int_{-\infty}^{\infty} 2\sqrt{\frac{2}{\pi}}\left(\frac{\sin as - as\cos as}{s^3}\right)\left(\cos sx-\sin sx\right)ds$$

$$=\frac{2}{\pi}\int_{-\infty}^{\infty}\left(\frac{\sin as - as\cos as}{s^3}\right)\cos sx\,dx - \frac{i2}{\pi}\left(\frac{\sin as - as\cos as}{s^3}\right)\sin sx\,ds$$

$$f(x)=\frac{4}{\pi}\int_{0}^{\infty}\left(\frac{\sin as - as\cos as}{s^3}\right)\cos sx\,dx - o$$

Put x = o

$$f(o)=\frac{4}{\pi}\int_{0}^{\infty}\frac{\sin as - as\cos as}{s^3}ds$$

$$=x\int_{0}^{\infty}\left(\frac{\sin as - as\cos as}{s^3}\right)ds=\frac{\pi}{4}f(o)=\frac{\pi}{4}a^2$$

Put $t = as \Rightarrow dt = ads$ limits (same)

$$\int_{0}^{\infty}\frac{\sin t - t\cos t}{\left(\frac{t^3}{a^3}\right)}\frac{dt}{a}=\frac{\pi}{4}a^2$$

$$\Rightarrow\int_{0}^{\infty}\left(\frac{\sin t - t\cos t}{t^3}\right)dt=\frac{\pi}{4}$$

3.4.1 Fourier Sine and Cosine Transforms

1. Let us discuss the Fourier cosine transform equation of e^{-ax}, a > o. Hence evaluate

$\int_{0}^{\infty}\frac{x^2}{(a^2+x^2)}dx$ and $\int_{0}^{\infty}\frac{x^2}{(a^2+x^2)(x^2+b^2)}$.

Solution:

$$F_s\left(e^{-ax}\right)=\sqrt{\frac{2}{\pi}}\int_{0}^{\pi}e^{-ax}\sin sx\,dx=\sqrt{\frac{2}{\pi}}\frac{s}{s^2+a^2}$$

$$F_c\left(e^{-ax}\right)=\sqrt{\frac{2}{\pi}}\int_{0}^{\infty}e^{-ax}\cos sx\,dx=\sqrt{\frac{2}{\pi}}\left(\frac{a}{s^2+a^2}\right)$$

By Parseval's identity:

$$\int_{0}^{\infty}F_s^2(s)ds=\int_{0}^{\infty}(f(x))^2dx$$

$$\Rightarrow \int_0^\infty \frac{2}{\pi}\left(\frac{s}{s^2+a^2}\right)^2 ds = \int_0^\infty e^{-2ax} dx$$

$$\Rightarrow \int_0^\infty \frac{x^2 dx}{\left(x^2+a^2\right)} = \frac{\pi}{4a}$$

Let $f(x)=e^{-ax};\ g(x)=e^{-bx}$

$$F_c(f(x))=\sqrt{\frac{2}{\pi}}\left(\frac{a}{s^2+a^2}\right), Fe(g/x)=\sqrt{\frac{2}{\pi}}\left(\frac{b}{s^2+b^2}\right)$$

$$\left(\therefore \int_0^\infty F_c(f(x))F_c(g(x))ds = \int_0^\infty f(x)g(x)dx\right)$$

$$\frac{2}{\pi}\int_0^\infty \frac{ab\,ds}{\left(a^2+s^2\right)\left(a^2+s^2\right)}\int_0^\infty e^{-(a+b)x} dx$$

$$\Rightarrow \int_0^\infty \frac{dx}{\left(a^2+x^2\right)\left(b^2+x^2\right)} = \frac{\pi}{(a+b)2ab.}$$

2. Let us solve the Fourier sine transform equation e^{-ax}, $a>0$.

Solution:

$$f\left(e^{-ax}\right)=\sqrt{\frac{2}{\pi}}\int_0^\infty e^{-ax}\sin sx\,dx$$

$$=\sqrt{\frac{2}{\pi}}\frac{a}{a^2+s^2}$$

$$f(x)=\sqrt{\frac{2}{\pi}}\int_0^\infty F_s(s)\sin sx\,ds$$

$$=\sqrt{\frac{2}{\pi}}\int_0^\infty \sqrt{\frac{2}{\pi}}\frac{a}{a^2+s^2}\sin sx\,ds$$

i.e. $$\int_0^\infty \frac{s\sin sx}{a^2+s^2} ds = \frac{\pi}{2} f(x)$$

i.e. $$\int_0^\infty \frac{s\sin sx}{a^2+s^2} ds = \frac{\pi}{2} e^{-ax}$$

3. Let us discuss the Fourier sine and cosine transform.

$$f(x)=\begin{cases}\sin x, & 0<x<a \\ 0, & x>a\end{cases}$$

Solution:

$$F_s[f(x)]=\sqrt{\frac{2}{\pi}}\int_0^\infty f(x)\sin sx\,dx$$

$$=\sqrt{\frac{2}{\pi}}\int_0^a \sin x \sin sx\,dx$$

$$=\sqrt{\frac{2}{\pi}}\int_0^a \frac{1}{2}\left[\sin(1+s)+\sin(1-s)x\right]dx$$

$$=\frac{1}{\sqrt{2\pi}}\left[\frac{\sin(1-s)x}{(1-s)}-\frac{\sin(1+s)x}{(1+s)}\right]_0^a$$

$$=\frac{1}{\sqrt{2\pi}}\left[\left(-\frac{\cos(1+s)a}{(1+s)}-\frac{\cos(1-s)a}{(1-s)}\right)-\left(-\frac{1}{1+s}-\frac{1}{1-s}\right)\right]$$

$$F_c[f(x)]=\frac{1}{\sqrt{2\pi}}\left[\left(\frac{1}{1+s}+\frac{1}{1-s}\right)-\left(\frac{\cos(1+s)a}{(1+s)}+\frac{\cos(1-s)a}{(1-s)}\right)\right]$$

4. Let us see an example of cosine transform method $\int_0^\infty \frac{dt}{(a^2+t^2)(b^2+t^2)}$.

Solution:

$$\text{Let } f(x)=e^{-ax};\ g(x)=e^{-bx} \qquad ...(1)$$

$$F_c\left[e^{-ax}\right]=\sqrt{\frac{2}{\pi}}\int_0^\infty e^{-ax}\cos sx\,dx$$

$$=\sqrt{\frac{2}{\pi}}\left[\left(\frac{e^{-ax}}{a^2+s^2}\right)(-a\cos sx+s\sin sx)\right]_0^\infty$$

$$=\sqrt{\frac{2}{\pi}}\left(\frac{a}{a^2+s^2}\right) \qquad ...(2)$$

$$\text{III}^{\text{ly}} F_c\left[e^{-bx}\right] = \sqrt{\frac{2}{\pi}}\left(\frac{b}{a^2+s^2}\right) \qquad \ldots(3)$$

WKT

$$\int_0^\infty F_c[f(x)].F_c[g(x)]\,ds = \int_0^\infty f(x).g(x)dx$$

$$\int_0^\infty \sqrt{\frac{2}{\pi}}\left(\frac{a}{a^2+s^2}\right).\sqrt{\frac{2}{\pi}}\left(\frac{b}{b^2+s^2}\right)ds = \int_0^\infty e^{-ax}.e^{-bx}dx$$

$$= \frac{2ab}{\pi}\int_0^\infty \frac{ds}{(a^2+s^2)(b^2+s^2)} = \int_0^\infty e^{-(a+b)x}dx$$

$$= \left[\frac{e^{-(a+b)x}}{-(a+b)}\right]_0^\infty = \frac{1}{(a+b)}$$

$$= \int_0^\infty \frac{ds}{(a^2+s^2)(b^2+s^2)} = \frac{\pi}{2ab(a+b)}$$

Put $s = t, ds = dt$

$$\therefore \int_0^\infty \frac{dt}{(a^2+t^2)(b^2+t^2)} = \frac{\pi}{2ab(a+b)}$$

5. Let us solve the cosine transform problem of $e^{-x^2/2}$.

Solution:

$$F_c(f(x)) = \sqrt{\frac{2}{\pi}}\int_0^\infty e^{-x^2/2}\cos sx\,dx$$

$$= \frac{1}{2}\text{Real Part of}\sqrt{\frac{2}{\pi}}\int_{-\infty}^\infty e^{-x^2/2}e^{-isx}dx$$

$$= \frac{\text{R.P}}{2}\text{of}\sqrt{\frac{2}{\pi}}\int_{-x}^\infty e^{-\frac{1}{2}(x-isx)^2\frac{s^2}{2}}dx$$

$$= e^{\frac{s^2}{2}}\frac{\text{RP of}}{2}\sqrt{\frac{2}{\pi}}\int_{-x}^\infty e^{-\frac{1}{2}(x-isx)^2}dx$$

$$= e^{-\frac{s^2}{2}}$$

6. Let us discuss the problem based on Fourier cosine transform $F_c e^{(-ax)}$, $F_c\left(\frac{1}{1+x^2}\right)$ and $F_s\left(\frac{1}{1+x^2}\right)$.

Solution:

$$F_c[f(x)] = \sqrt{\frac{2}{\pi}}\int_0^\infty f(x)\cos sx\,dx$$

$$F_c\left[e^{-ax}\right] = \sqrt{\frac{2}{\pi}}\int_0^\infty e^{-ax}\cos sx\,dx$$

$$= \sqrt{\frac{2}{\pi}}\left[\frac{e^{-ax}}{a^2+s^2}(-a\cos sx + s\sin sx)\right]_0^\infty$$

$$F_c\left[e^{-ax}\right] = \sqrt{\frac{2}{\pi}}\left(\frac{a}{s^2+a^2}\right) \quad ...(1)$$

$$F_c\left(\frac{1}{1+x^2}\right) = \sqrt{\frac{2}{\pi}}\int_0^\infty\left(\frac{1}{1+x^2}\right)\cos sx\,dx \quad ...(2)$$

WKT $F_c\left(e^{-x}\right) = \sqrt{\frac{2}{\pi}}\left(\frac{1}{s^2+1}\right)$

Using inversion formula:

$$f(x) = \sqrt{\frac{2}{\pi}}\int_0^\infty F_c[f(x)]\cos sx\,ds$$

$$e^{-x} = \sqrt{\frac{2}{\pi}}\int_0^\infty\sqrt{\frac{2}{\pi}}\left(\frac{1}{1+s^2}\right)\cos sx\,ds$$

i.e. $= \sqrt{\frac{2}{\pi}}\int_0^\infty\frac{1}{1+s^2}\cos sx\,ds\sqrt{\frac{\pi}{2}}e^{-x}$

Put s = x

$$\sqrt{\frac{2}{\pi}}\int_0^\infty\frac{1}{1+x^2}\cos sx\,dx = \sqrt{\frac{2}{\pi}}e^{-s}$$

i.e., $F_c\left[\frac{1}{1+x^2}\right]=\sqrt{\frac{\pi}{2}}e^{-s}\,\text{by}(2)$

WKT $F_s\left[e^{-x}\right]=\sqrt{\frac{2}{\pi}}\left(\frac{s}{s^2+1}\right)$

$$F_s\left[\frac{x}{1+x^2}\right]=\sqrt{\frac{2}{\pi}}\int_0^\infty\frac{x}{1+x^2}\sin sx\,dx \qquad ...(3)$$

By inversion formula:

$$f(x)=\sqrt{\frac{2}{\pi}}\int_0^\infty F_s\left[f(x)\right]\sin sx\,ds$$

$$e^{-x}=\sqrt{\frac{2}{\pi}}\int_0^\infty\sqrt{\frac{2}{\pi}}\left(\frac{s}{s^2+1}\right)\sin sx\,dx$$

$$\sqrt{\frac{2}{\pi}}\int_0^\infty\left(\frac{s}{s^2+2}\right)\sin sx\,ds=\sqrt{\frac{\pi}{2}}e^{-x}$$

Put s = x

$$F_s\left[\frac{x}{1+x^2}\right]=\sqrt{\frac{\pi}{2}}e^{-s}$$

7. Let us solve the cosine transform of $e^{-a^2x^2}$ for any a > 0 and hence prove that $e^{-x^2/2}$ is self-reciprocal under Fourier Cosine transform.

Solution:

We Know That Fourier Cosine Transform of f(x) is:

$$F_c\left[f(x)\right]=\sqrt{\frac{2}{\pi}}\int_0^\infty f(x)\cos sx\,dx$$

$$=\sqrt{\frac{2}{\pi}}\int_0^\infty e^{-a^2x^2}\cos sx\,dx$$

$$=\frac{1}{2}\sqrt{\frac{2}{\pi}}\int_{-\infty}^\infty e^{-a^2x^2}\cos sx\,dx \qquad \left(\because e^{-a^2x^2}\cos sx\,\text{is even}\right)$$

$$=\frac{1}{\sqrt{2\pi}}\int_{-\infty}^\infty e^{-a^2x^2}.\text{R.P of } e^{isx}dx$$

$$= R.P of \frac{1}{\sqrt{2\pi}} \int_{-\infty}^{\infty} e^{-a^2} \left[x^2 - \frac{isx}{a^2} \right] dx \qquad ...(1)$$

Now consider:

$$x^2 - \frac{isx}{a^2} = x^2 - \frac{isx}{a^2} + \left(\frac{is}{2a^2}\right) - \left(\frac{is}{2a^2}\right)^2$$

$$= \left(x - \frac{is}{2a^2} \right)^2 + \frac{S^2}{4a^4}$$

Substituting in (1):

$$F_c[f(x)] = R.P of \int_{-\infty}^{\infty} e^{-a^2} \left[\left(x - \frac{is}{2a^2} \right) + \frac{S^2}{4a^4} \right]_{dx}$$

$$= R.P of \frac{1}{\sqrt{2\pi}} e^{-\frac{s^2}{4a^2}} \int_{-\infty}^{\infty} e^{-\left(ax - \frac{is}{2a}\right)^2} dx$$

Put $ax - \dfrac{is}{2a} = u \Rightarrow a\,dx = du$

$$F_c[f(x)] = R.P of \frac{1}{\sqrt{2\pi}} e^{-\frac{s^2}{4a^2}} \int_{-\infty}^{\infty} e^{-u^2} \frac{du}{a}$$

$$= R.P of \frac{1}{a\sqrt{2\pi}} e^{-\frac{s^2}{4a^2}} \int_{-\infty}^{\infty} e^{-u^2} du$$

$$= R.P of \frac{1}{a\sqrt{2\pi}} e^{-\frac{s^2}{4a^2}} . \sqrt{\pi} \left(\because \int_{-\infty}^{\infty} e^{-x^2} x = \sqrt{\pi} \right)$$

$$= R.P of \frac{1}{a\sqrt{2\pi}} e^{-\frac{s^2}{4a^2}}$$

Hence $F_c\left[e^{-a^2x^2}\right] = \dfrac{1}{a\sqrt{2}} e^{-\frac{s^2}{4a^2}}$ ...(2)

Put,

$$a^2 = \frac{1}{2} \text{ i.e., } a = \frac{1}{\sqrt{2}} \text{ in } (2)$$

$$F_c\left[e^{-\frac{x^2}{2}}\right]=e^{-\frac{s^2}{2}}$$

Hence $e^{-x^2/2}$ is self-reciprocal under Fourier Cosine transform.

8. Let us see an example of cosine transform.

Solution:

$$\text{WKT, F}\left[f(x)\right]\ \sqrt{-}\int^{\infty} f(x)\cos sx\,dx$$

$$\text{Now } F_c\left[e^{-a^2x^2}\right]=\sqrt{\frac{2}{\pi}}\int_0^{\infty} e^{-a^2x^2}\cos sx\,dx$$

$$=\frac{1}{2}\sqrt{\frac{2}{\pi}}\int_{-\infty}^{\infty} e^{-a^2x^2}\cos Sx\,dx$$

$$\left(\because e^{-a^2x^2}\cos Sx \text{ is even function}\right)$$

$$=\frac{1}{\sqrt{2\pi}}\int_{-\infty}^{\infty} e^{-a^2}\text{Real part of } e^{isx}dx$$

$$=\text{R.P of}\frac{1}{\sqrt{2\pi}}\int_{-\infty}^{\infty} e^{-a^2}\left[x^2-\frac{isx}{a^2}\right]dx \qquad ...(1)$$

$$\text{Now}\quad x^2-\frac{isx}{a^2}=x^2-\frac{isx}{a^2}+\left(\frac{is^2}{2a^2}\right)-\left(\frac{is^2}{2a^2}\right)$$

$$=\left(x-\frac{is^2}{2a^2}\right)^2+\frac{s^2}{4a^4}$$

$$=e^{-a^2x^2}e^{-a^2x^2}$$

Substituting (1):

$$F_C\left[e^{-a^2x^2}\right]=\text{R.P of}\frac{1}{\sqrt{2\pi}}\int_{-\infty}^{\infty} e^{-a^2}\left[\left(x-\frac{is}{2a^2}\right)^2+\frac{S^2}{4a^4}\right]dx$$

$$=\text{R.P of}\frac{1}{\sqrt{2\pi}}e^{-\frac{s^2}{4a^2}}\int_{-\infty}^{\infty} e^{-}\left(ax-\frac{is}{2a}\right)^2 dx$$

Put $ax-\frac{is}{2a}=u \Rightarrow adx = du$

$$F_C\left[f(x)\right]=\text{R.P of}\frac{1}{\sqrt{2\pi}}e^{-s^2/4a^2}\int_{-\infty}^{\pi} e-u^2\frac{du}{a}$$

Limit:

$$x \to \infty \Rightarrow u \to \infty$$

$$x \to -\infty \Rightarrow u \to -\infty$$

$$=\text{R.P of}\frac{1}{a\sqrt{2\pi}}.e^{-\frac{s^2}{4a^2}}\int_{-\infty}^{\infty} e^{-u^2}du$$

$$=\text{R.P of}\frac{1}{a\sqrt{2\pi}}e^{-\frac{s^2}{4a^2}}.\sqrt{\pi}\quad\left(\because\int_{-\infty}^{\infty} e^{-x^2}dx=\sqrt{\pi}\right)$$

$$\therefore F_C\left[e^{-a^2x^2}\right]=\frac{1}{a\sqrt{2}}e^{\frac{-s^2}{4a^2}} \qquad \text{...(2)}$$

WKT, $F_s\left[xf(x)\right]=\frac{-d}{ds}F_C\left[f(x)\right]$

$$\therefore F_S\left[xe^{-a^2x^2}\right]=\frac{-d}{ds}\frac{1}{a\sqrt{2}}e^{-\frac{s^2}{4a^2}}$$

$$=\frac{-1}{a\sqrt{2}}\left(\frac{-2s}{4a^2}\right)e^{\frac{-s^2}{4a^2}}$$

$$=\frac{+s}{a^3 2\sqrt{2}}e^{-\frac{s^2}{4a^2}}$$

$$=\frac{s}{a^3 2\sqrt{2}}e^{\frac{-s^2}{4a^2}}$$

9. Let us solve the Fourier sine transform for the given problem $f(x)=\frac{e^{-ax}}{x}$.

Solution:

WKT Fourier sine transform of e^{-ax} is

$$F_s\left[e^{-ax}\right]=\sqrt{\frac{2}{\pi}}\left(\frac{s}{s^2+a^2}\right)$$

By definition $\sqrt{\frac{2}{\pi}}\int_0^\infty e^{-ax}\sin sx\,dx=\sqrt{\frac{2}{\pi}}\left(\frac{s}{s^2+a^2}\right)$

i.e., Integrating both side with respect to 'a' we get:

$$\sqrt{\frac{2}{\pi}}\int_0^\infty \frac{e^{-ax}}{-x}\sin sx\,dx=\sqrt{\frac{2}{\pi}}\int\left(\frac{s}{s^2+a^2}\right)da$$

(In the left side, we have integrated e^{-ax} with respect to 'a' as a x are independent):

$$=s\sqrt{\frac{2}{\pi}}\frac{1}{s}\tan^{-1}\left(\frac{a}{s}\right)=\sqrt{\frac{2}{\pi}}\tan^{-1}\frac{a}{s}$$

$$\text{i.e.,}\sqrt{\frac{2}{\pi}}\int_0^\infty\left(\frac{e^{-ax}}{x}\right)\sin sx\,dx=-\sqrt{\frac{2}{\pi}}\tan^{-1}\frac{a}{s}$$

$$\text{i.e.,}F_S\left[+\frac{e^{-ax}}{x}\right]=-\sqrt{\frac{2}{\pi}}\tan^{-1}\left(\frac{a}{s}\right)$$

10. Let us solve sine and cosine transform of e^{-ax}, a> fl hence evaluate $\int_0^\infty \frac{dx}{\left(a^2+x^2\right)\left(b^2+x^2\right)}$.

Solution:

Let, $f(x)=e^{-ax}$

WKT, $F_c\left[f(x)\right]=\sqrt{\frac{2}{\pi}}\int_0^\infty f(x)\cos sx\,dx$

(i) $F\left[e^{-ax}\right]=\sqrt{\frac{2}{\pi}}\int_0^\infty e^{-ax}\cos sxdx$

$$=\sqrt{\frac{2}{\pi}}\left[\frac{e^{-ax}}{a^2+s^2}(-a\cos sx+s\sin sx)\right]_0^\infty$$

$$F_c\left[e^{-ax}\right]=\sqrt{\frac{2}{\pi}}\left(\frac{a}{a^2+s^2}\right).$$

(ii) WKT, $F_s[f(x)] = \sqrt{\frac{2}{\pi}}\int_0^\infty e^{-ax}\sin sx\,dx$

$$= \sqrt{\frac{2}{\pi}}\int_0^\infty e^{-ax}\sin sx\,dx$$

$$= \sqrt{\frac{2}{\pi}}\left[\frac{e^{-ax}}{a^2+s^2}(-a\sin sx - s\cos sx)\right]_0^\infty$$

$$F_s\left[e^{-ax}\right] = \sqrt{\frac{2}{\pi}}\left(\frac{s}{a^2+s^2}\right)$$

Let $f(x) = e^{-ax};\ g(x) = e^{-bx}$

WKT:

$$F_c\left[e^{-ax}\right] = \sqrt{\frac{2}{\pi}}\left(\frac{a}{a^2+s^2}\right)$$

$$F_c\left[e^{-bx}\right] = \sqrt{\frac{2}{\pi}}\left(\frac{b}{b^2+s^2}\right)$$

By Parseval's identity:

$$\int_0^\infty F_c[f(x)]F_c[g(x)]ds = \int_0^\infty f(x).g(x)dx$$

$$= \int_0^\infty \sqrt{\frac{2}{\pi}}\left(\frac{a}{a^2+s^2}\right)\sqrt{\frac{2}{\pi}}\left(\frac{b}{b^2+s^2}\right)ds = \int_0^\infty e^{-ax}.e^{-bx}dx$$

$$= \frac{2ab}{\pi}\int_0^\infty \frac{ds}{(a^2+s^2)(b^2+s^2)} = \int_0^\infty e^{-(a+b)x}dx$$

$$= \left[\frac{e^{-(a+b)}}{-(a+b)}\right]_0^\infty = \frac{1}{(a+b)}$$

$$\text{i.e.,} \int_0^\infty \frac{ds}{(a^2+s^2)(b^2+s^2)} = \frac{\pi}{2ab(a+b)}.$$

11. Let us solve the problem based on cosine transform $\int_0^\infty F(x)g(x)dx = \int_0^\infty F_C(s)G_C(s)ds$.

Solution:

We Know that:

$$F_c[f(x)] = \sqrt{\frac{2}{\pi}}\int_0^\infty f(x)\cos sx\,dx = F_C(s) \quad ...(1)$$

$$G_c[g(x)] = \sqrt{\frac{2}{\pi}}\int_0^\infty g(x)\cos sx\,dx = G_C(s) \quad ...(2)$$

and inverse formula: $f(x) = \sqrt{\frac{2}{\pi}}\int_0^\infty F_c(s)\cos sx\,ds$ ...(3)

$$g(x) = \sqrt{\frac{2}{\pi}}\int_0^\infty G_c(s)\cos sx\,dx \quad ...(4)$$

Consider RHS:

$$\int_0^\infty F_c(s).G_c(s)ds = \int_0^\infty F_c(s)\sqrt{\frac{2}{\pi}}\int_0^\infty g(x)\cos sx\,dx\,ds$$

$$= \int_0^\infty g(x)\left[\sqrt{\frac{2}{\pi}}\int_0^\infty F_c(s)\cos sx\,dx\right]ds$$

$$= \int_0^\infty g(x)f(x)dx$$

Here, $\int_0^\infty f(x)g(x)dx = \int_0^\infty F_c(s)G_c(s)ds$

12. Let us solve a problem based on cosine transform e^{-x^2}.

Solution:

$$\text{Let } f(x) = e^{-x^2}$$

WKT, $F_c[f(x)] = \sqrt{\frac{2}{\pi}}\int_0^\infty f(x)\cos sx\,dx$

$$= \sqrt{\frac{2}{\pi}}\int_0^\infty e^{-x^2}\cos sx\,dx$$

$$=\frac{1}{\sqrt{2\pi}}\int_0^{\infty} e^{-x^2}\cos sx\,dx\left[\because \int_{-\infty}^{\infty} I(x)dx = 2\int_0^{\infty} I(x)dx\right]$$

If I(x) is even:

$$=R.P\,of\int_0^{\infty} e^{-x^2}e^{isx}dx\left[\because \cos sx = R.P\,of\,e^{isx}\right]$$

$$=R.P\,of\frac{1}{\sqrt{2\pi}}\int_{-\infty}^{\infty} e^{-\left(x^2-isx\right)}dx$$

$$\left[Now\,x^2-isx = x^2-isx+\left(\frac{is}{2}\right)^2-\left(\frac{is}{2}\right)^2=\left(x-\frac{is}{2}\right)^2+\frac{s^2}{4}\right]$$

$$\therefore F_c\left[f(x)\right]=R.P\,of\frac{1}{\sqrt{2\pi}}\int_{-\infty}^{\infty} e^{-\left[\left(x-\frac{is}{2}\right)^2+\frac{s^2}{4}\right]}dx$$

$$=R.P\;of\;\frac{e^{-\frac{s^2}{4}}}{\sqrt{2\pi}}\int_{-\infty}^{\infty} e^{-\left(x-\frac{is}{2}\right)^2}dx$$

Put $t = x-\frac{is}{2} \Rightarrow dt = dx$

$x\to-\infty \quad t\to-\infty$

$x\to\infty \quad t\to\infty$

$$=F_c\left[f(x)\right]=R.P\,of\frac{-s^2/4}{\sqrt{2\pi}}\int_{-\infty}^{\infty} e^{-t^2}dt$$

$$=R.P\,of\frac{e^{-s^2/4}}{\sqrt{2\pi}}\sqrt{\pi}\left(\because \int_{-\infty}^{\infty} e^{-t^2}du=\sqrt{\pi}\right)$$

$$F_c\left[e^{-x^2}\right]=\frac{1}{\sqrt{2}}e^{-s^2/4}$$

13. Let us discuss the Fourier transform $\frac{1}{\sqrt{x}}$.

Solution:

Consider:

$$\sqrt{\frac{2}{\pi}}\int_{o}^{\infty} x^{n-1}e^{-isx}$$

Put, $t = isx$

$dt = is\,dx$

$x \to o \quad t \to o$
$x \to \infty \quad t \to \infty$

$$= \sqrt{\frac{2}{\pi}}\int_{o}^{\infty} x^{n-1}e^{-isx}dx = \sqrt{\frac{2}{\pi}}\int_{o}^{\infty} \frac{t^{n-1}}{(i)^{n-1}(s)^{n-1}}e^{-t}\frac{dt}{is}$$

$$= \sqrt{\frac{2}{\pi}}\frac{1}{(i)^{n}(s)^{n}}\int_{o}^{\infty} e^{-t}e^{n-1}dt$$

$$\left[\text{WKT}\frac{1}{(i)^{n}} = (-i)^{n} = \cos\frac{n\pi}{2}i\ \sin\frac{n\pi}{2}\overline{|(n)} = \int_{o}^{\infty} e^{-x}x^{n-1}dx\right]$$

$$\therefore \sqrt{\frac{2}{\pi}}\int_{o}^{\infty} x^{n-1}e^{-isx}dx = \sqrt{\frac{2}{\pi}}\frac{(-i)^{n}}{s^{n}}\overline{|(n)}$$

$$\text{i.e.,}\sqrt{\frac{2}{\pi}}\int_{o}^{\infty} x^{n-1}(\cos sx - i\sin sx)dx$$

$$\sqrt{\frac{2}{\pi}}\frac{\overline{|(n)}}{s^{n}}\left(\cos\frac{n\pi}{2} - i\sin\frac{n\pi}{2}\right)$$

$$\text{i.e.,}\sqrt{\frac{2}{\pi}}\int_{o}^{\infty} x^{n-1}\cos sx\,dx - i\sqrt{\frac{2}{\pi}}\int_{o}^{\infty} x^{n-1}\sin sx\,dx$$

$$Fc\left[x^{n-1}\right] - iF_{s}\left[x^{n-1}\right] = \sqrt{\frac{2}{\pi}}\frac{\overline{|(n)}}{s^{n}}\cos\frac{n\pi}{2}$$

$$= -i\sqrt{\frac{2}{\pi}}\frac{\overline{|(n)}}{s^{n}}\sin\frac{n\pi}{2}$$

Equaling Real and imaginary parts:

$$F_c\left[x^{n-1}\right]=\sqrt{\frac{2}{\pi}}\frac{\lceil(n)}{s^n}\cos\frac{n\pi}{2} \quad \text{...(1)}$$

$$F_s\left[x^{n-1}\right]=\sqrt{\frac{2}{\pi}}\frac{\lceil(n)}{s^n}\sin\frac{n\pi}{2} \quad \text{...(2)}$$

Put n = 1/2 in (1) and (2):

$$F_c\left[x^{-1/2}\right]=\sqrt{\frac{2}{\pi}}\frac{\lceil(1/2)}{s^{1/2}}\cos\frac{\pi}{4}=\sqrt{\cdot\frac{2}{\pi}}\frac{\sqrt{\pi}}{s^{1/2}}\frac{1}{\sqrt{2}}$$

$$\text{i.e.,}\left[\frac{1}{x^{1/2}}\right]=\frac{1}{s^{1/2}} \qquad \left(\because \lceil(1/2)=\sqrt{\pi}\right).$$

Hence $\frac{1}{\sqrt{x}}$ is self-reciprocal under Fourier Cosine transform.

Put n =1/2 in (2).

$$F_S\left[\frac{1}{\sqrt{x}}\right]=\sqrt{\frac{2}{\pi}}\frac{\lceil(1/2)}{s^{1/2}}\sin\frac{\pi}{4}$$

$$=\sqrt{\frac{2}{\pi}}\frac{\sqrt{\pi}}{s^{1/2}}\frac{1}{\sqrt{2}}$$

$$=\frac{1}{\sqrt{s}}$$

Hence $\frac{1}{\sqrt{x}}$ is self-reciprocal under Fourier sine transform.

14. Let us see an example of sine transform.

$$f(x)\begin{cases} x & , \quad 0<x<1 \\ 2-x & , \quad 1<x<2 \\ 0 & , \quad x>2. \end{cases}$$

Solution:

Fourier Sine Transform of f(x):

$$F_s\left[f(x)\right]=\sqrt{\frac{2}{\pi}}\int_0^{\infty}f(x)\sin sx\,dx$$

$$=\sqrt{\frac{2}{\pi}}\left[\int_0^1 x\sin sx\,dx+\int_1^2(2-x)\sin sx\,dx+0\right]$$

$$=\sqrt{\frac{2}{\pi}}\left[\left\{(x)\left(-\frac{\cos sx}{s}\right)-(-1)\left(-\frac{\sin sx}{s^2}\right)\right\}_0^1\right.$$

$$\left.+\left\{(2-x)\left(-\frac{\cos sx}{s}\right)-(-1)\left(-\frac{\sin sx}{s^2}\right)\right\}_1^2\right]$$

$$=\sqrt{\frac{2}{\pi}}\left[\left\{\left(-\frac{\cos s}{s}+\frac{\sin s}{s^2}\right)-(0)\right\}+\left\{\left(-\frac{\sin 2s}{s^2}\right)-\left(-\frac{\cos s}{s}-\frac{\sin s}{s^2}\right)\right\}\right]$$

$$=\sqrt{\frac{2}{\pi}}\left[-\frac{\cos s}{s}+\frac{\sin s}{s^2}-\frac{\sin 2s}{s^2}+\frac{\cos s}{s}+\frac{\sin s}{s^2}\right]$$

$$F_s\left[f(x)\right]=\sqrt{\frac{2}{\pi}}\left[\frac{2\sin s-\sin 2s}{s^2}\right].$$

3.4.2 Properties

1. Linearity Property

If $F_1(s)$ and $F_2(s)$ be the Fourier transform $f_1(x)$ and $f_2(x)$ respectively, then $F\left[af_1(x)+bf_2(x)\right]=aF_1+bF_2(s)$ where a and b are contents.

Proof:

$$\text{L.H.S}=\int_{-\infty}^{\infty}\left[af_1(t)+bf_2(t)\right].e^{ist}\,dt$$

$$=a\int_{-\infty}^{\infty}f_1(t).e^{ist}dt+b\int_{-\infty}^{\infty}f_2(t).e^{ist}\,dt$$

$$=aF_1(s)+bF_2(s)=\text{R.H.S.}$$

2. Change of Scale Property

$$\text{If } F\left[f(t)\right]=F(s),\text{then } F\left[f(at)\right]=\frac{1}{a}F(s/a),a>0$$

Proof:

$$F\left[f(at)\right]=\int_{-\infty}^{\infty}f(at).e^{ist}dt$$

$$=\frac{1}{a}\int_{-\infty}^{\infty}f(p).e^{i\frac{1}{a}p}dp \qquad \text{(Taking at = p and adt = dp)}$$

$$=\frac{1}{a}.F(s/a)$$

From this change of scale property it is evident that if the width of a function is decreased while its height is kept constant, then its Fourier transform wider and shorter. If its width is increased, its transform becomes narrower and taller.

A similar frequency scaling property is given by:

$$F\left[\frac{1}{a}.f(t/a)\right]=F(as)$$

3. Shifting Property

If $F\{f(t)\}=F(s)$ and a is a real constant then, $F\{f(t-a)\}=e^{ixa}.F(s)$

Proof:

$$F\{f(t-a)\}=\int_{-\infty}^{\infty}f(t-a)e^{ixt}dt$$

$$=e^{isa}\int_{-\infty}^{\infty}f(t-a)e^{is(t-a)}.dt$$

$$=e^{isa}\int_{-\infty}^{\infty}f(p)e^{isp}\,dp \qquad (\text{taking}\, t-a=p\, \text{and}\, dt=dp)$$

$$=e^{isa}.F(s).$$

This (time) shifting property states that the Fourier transform of a shifted function is just the transform of the unshifted function multiplied by an exponential factor having a linear phase. Likewise, the frequency shifting property states that if F(s) is shifted by a constant 'a', its inverse transform is multiplied by e^{-ita}.

$$F\{f(t).e^{-ita}\}=F\{s-a\}$$

4. Modulation Property

If $F\{f(t)\} = F(s)$, then $F[f(t)\cos wt] = \frac{1}{3}[F(s-w)+F(s+w)]$

Proof:

$$F[f(t).\cos wt] = \int_{-\infty}^{\infty} f(t).\cos wt.e^{ixt}.dt$$

$$= \int_{-\infty}^{\infty} f(t).\left(\frac{e^{iwt}+e^{iwt}}{2}\right).e^{ixt}.dt$$

$$= \frac{1}{2}\left[\int_{-\infty}^{\infty} f(t).e^{i(x+w)t}dt + \int_{-\infty}^{\infty} f(t).e^{i(s-w)t}dt\right]$$

$$= \frac{1}{2}[F(s+w)+F(s-w)]$$

This result has application in radio and television where the harmonic carrier wave is modulated by an envelope.

5. Convolution Property

The convolution of two function f(t) and g(t) in $(-\infty,\infty)$ is defined as:

$$f(t)^{*} g(t) = \int_{-\infty}^{\infty} f(t).g(x-t)dt].$$

Let, $F\{f(t)\} = F_1(s)$ and $F\{g(t)\} = F_2(s)$, then:

$$F[f(t)*g(t)] = F_1(s).\ F_2(s)$$

Proof:

$$F\{f(t)^{*} g(t)\} = \int_{-\infty}\{f(t)^{*} g(t)\}.e^{ixt}dt$$

$$= \int_{-\infty}^{\infty}\left(\int_{-\infty}^{\infty} f(u)g(t-u)du\right)e^{ixt}dt$$

$$= \int_{-\infty}^{\infty} f(u)e^{ixt}\left(\int_{-\infty}^{\infty} g(t-u).e^{is(t-U)}d(t-u)\right)du$$

On changing the order of integration:

$$= \int_{-\infty}^{\infty} f(u).e^{ixu}.F_2(s)du = \left(\int_{-\infty}^{\infty} f(u).e^{ixu}du\right).F_2(s)$$

$$=F_1(s).F_2(s)$$

6. Correlation Property

The correlation integral, like the convolution integral is important in theoretical and practical applications.

The correlation integral is defined as, $h(t)=\int_{-\infty}^{\infty} f(u)g(t+u)du$ and like the convolution property, it forms a transform pair given by:

$$F\{h(t)\}=F(s).G^*(s).$$

If f(t) and g(t) are the same function, the integral above is normally called the autocorrelation function and the cross correlation if they differ. The Fourier transform pair for the auto correlation is simply:

$$F\left\{\int_{-\infty}^{\infty} f(u).f(t+u)du\right\}=|F(s)|^2$$

The symmetry properties of the Fourier transform are given in the following table.

3.4.3 Inverse Transforms

A function $F(\omega)$ is called the Fourier transform of f(x), if

$$F(\omega):=F\{f(t)\}:=\int_{-\infty}^{\infty} e^{-i\omega t}f(t)dt \text{ exist.}$$

$$F^{-1}\{F(\omega)\}:\frac{1}{2\pi}\int_{-\infty}^{\infty} e^{i\omega x}F(\omega)d\omega \text{ is called the inverse Fourier transform of } F(\omega).$$

The Fourier transform of f is therefore a function $F\{f(t)\}$ of the new variable ω. This function, evaluated at ω, is $F(\omega)$.

The constants 1 and $1/2\pi$ preceding the integral signs could be replaced by any two constants whose product is $1/2\pi$.

The Fourier transform of f given by:

$$f(t):=\begin{cases}1 & |t|<1\\ 0 & |t|>1\end{cases}$$

$$F\{f(t)\}=\int_{-1}^{1} e^{-iwt}dt=\begin{cases}2\dfrac{\sin\omega}{\omega} & \omega\neq 0\\ 2 & \omega=0\end{cases}$$

The inverse Fourier transforms computes to:

$$\frac{1}{\pi}\int_{-\infty}^{\infty} e^{iwx}\frac{\sin\omega}{\omega}d\omega = \frac{2}{\pi}\int_{0}^{\infty}\cos\omega x\frac{\sin\omega}{\omega}d\omega$$

$$= \frac{1}{\pi}\int_{0}^{\infty}\frac{\sin\omega(x+1)}{\omega}d\omega - \frac{1}{\pi}\int_{0}^{\infty}\frac{\sin\omega(x-1)}{\omega}d\omega$$

$$= \begin{cases} 1 & |x|<1 \\ 1/2 & |x|=1 \\ 0 & |x|>1 \end{cases}$$

3.4.4 Finite Fourier Transforms

If the interval is finite interval $0 < x < L$ then depending on the boundary conditions we use finite Fourier sine or cosine transforms.

a) Finite Fourier Sine Transforms

$$\bar{u}_s(n,t) = F_s[u] = \int_0^1 u(x,t)\sin\left(\frac{n\pi x}{L}\right)dx$$

and the inverse sine transform is:

$$u(x,t) = \frac{2}{L}\sum_{n=1}^{x}\bar{u}_s(n,t)\sin\frac{n\pi x}{L}$$

b) Finite Fourier Cosine Transforms

$$\bar{u}_c(n,t) = F_c[u] = \int_0^L u(x,t)\cos\frac{n\pi x}{L}dx$$

and the inverse cosine transform is:

$$u(x,t) = \frac{1}{L}\bar{u}_c(0,t) + \frac{2}{L}\sum_{n=1}^{\infty}\bar{u}_c(n,t)\cos\frac{n\pi x}{L}$$

- If the interval is $0 < x < L$ and i) If boundary conditions are $u(0, t) = u(L, t) = 0$ for all t, we use finite Fourier sine transform.
- If boundary conditions are $u_x(0, t) = u_x(L, t) = 0$ for all t, we use finite Fourier cosine transform.

Interval	Boundary conditions	Type of F.T
$0 < x < L$	$u(0,t) = u(L,t) = 0$	Finite Fourier sine transform
$0 < x < L$	$\frac{\partial u}{\partial x}(0,t) = \frac{\partial u}{\partial x}(L,t) = 0$	Finite Fourier cosine transform

Example Problem on Finite Fourier Sine Transform

1. Let us use finite Fourier transform to solve $\frac{\partial u}{\partial x} - \frac{\partial^2 u}{\partial x^2} 0 < x < \pi, t > 0$s subject to:

- $u(x,0) = 1 \quad 0 < x < \pi$
- $u(0,t) = u(r,t) = 0, \quad 0 < x < \pi, \ t > 0.$

Solution:

Step 1: As u(0, t) = u(n, Let (Π, t) = 0 since we use finite Fourier sine transform.

$$\bar{u}_s(n,t) = F_s(u) = \int_0^L u(x,t)\sin\frac{n\pi x}{L}dx$$

$$L = \pi$$

$$\bar{u}_s(n,t) = F_s(u) = \int_0^\pi u(x,t)\sin nx\,dx$$

Step 2: Taking finite Fourier sine transform of $\frac{\partial u}{\partial t} = \frac{\partial^2 u}{\partial x^2}$ in $0 < x < \pi$ we get:

$$\int_0^\pi \frac{\partial u}{\partial t}\sin nx\,dx = \int_0^\pi \frac{\partial^2 u}{\partial x^2}\sin nx\,dx$$

Step 3: We use D.U.I.S on L.H.S and integration by parts on R.H.S.

$$\frac{d}{dt}\int_0^\pi u\sin nx\,dx = \left\{\left(\frac{\partial u}{\partial x}\sin nx\right)_0^\pi - \int_0^\pi \frac{\partial u}{\partial x}.n\cos nx\,dx\right\}$$

$$\frac{d}{dt}\bar{u}_s(n,t) = \left\{(0-0) - n\int_0^\pi \frac{\partial u}{\partial x}\cos nx\,dx\right\}$$

Apply integration by parts again:

$$\frac{d}{dt}\bar{u}_s(n,t) = (-n)\left\{(u.\cos nx)_0^\pi - \int_0^\pi u(-n)\sin nx\,dx\right\}$$

$$=(-n)\{(0-0)+n\bar{u}_s(n,t)\}$$

$$=-n^2.\bar{u}_s(n,t)$$

Step 4:

$$\frac{\frac{d}{dt}\bar{u}_s(n,t)}{\bar{u}_s(n,t)}=-n^2$$

By integrating with respect to t we get:

$$\log\bar{u}_s(n,t)=-n^2t+\log A$$

$$\therefore\ \bar{u}_s(n,t)=Ae^{-n^2t}$$

Step 5:

Put t=0

$$\bar{u},(n,0)=A$$

Step 6:

Put t = 0 in (i)

$$\bar{u},(n,0)=\int_0^{\pi}u(x,0)\sin nx\,dx$$

$$A=\int_0^{\pi}(1)\sin nx\,dx$$

$$=\left[-\frac{\cos nx}{n}\right]_0^{\pi}$$

$$=-\frac{1}{n}[\cos n\pi-1]$$

$$A=\frac{1}{n}\left[1-(-1)^n\right]$$

Step 7: substituting we get:

$$\bar{u}_s,(n,t)=\frac{\left[1-(-1)^n\right]}{n}e^{-n^2t}$$

Step 8:

$$\bar{u},(x,t)=\frac{2}{L}\sum_{n=1}^{\infty}\bar{u}_s(n,t)\sin\frac{n\pi x}{L}$$

$$L=\pi$$

$$u(x,t)=\frac{2}{\pi}\sum_{n=1}^{\infty}\frac{\left[1-(-1)^n\right]}{n}e^{-n^2t}.\sin nx$$

2. Let us calculate the finite Fourier (a) sine and b) cosine transforms of f(x) =x in 0<x<Π.

Solution:

(a) The finite Fourier sine transforms f(x) =x in 0<x<l is:

$$F_x(n)=\int_0^1 f(x)\sin\frac{n\pi x}{1}dx=\int_0^x x\sin nx\,dx$$

$$(\therefore f(x)=x,1=\pi)$$

$$\left[x\left(-\frac{\cos nx}{n}\right)-1\left(-\frac{\sin nx}{x^2}\right)\right]_0^x=-\frac{\cos n\pi}{n}=\frac{(-1)^{n-1}}{n}$$

$$=0\,at\,0 \qquad =0\,at\,\pi,0$$

(b)The finite Fourier cosine transforms f(x) =x in 0<x<l is:

$$F_c(n)=\int_0^1 f(x)\cos\frac{n\pi x}{1}dx;\text{ Here } 1=\pi \text{ and } f(x)=x.$$

So,

$$F_c(n)=\int_0^{\pi}x\cos nx\,dx=\left[x\frac{\sin nx}{n}-1.\left(-\frac{\cos nx}{n^2}\right)\right]_0^{\pi}$$

$$=\frac{\cos n\pi}{n^2}=\frac{(-1)^{n-1}}{n^2}=\begin{cases}0 & \text{if } n \text{ is even;}\\ \frac{-2}{n^2} & \text{if } n \text{ is odd}\end{cases}$$

3. Let us calculate the finite Fourier (a) sine and b) cosine transforms of

$$f(x)=\begin{cases}x & \text{if } 0\le x\ \pi/2\\ \pi-x & \text{if } 0\ \pi/2\le x\ \pi\end{cases}$$

Solution:

(a) The finite Fourier sine transforms:

$$F_s(n)=\int_0^{\pi} f(x)\sin nx\,dx=\int_0^{\pi/2} x\sin nx\,dx$$

$$+\int_0^{\pi/2}(\pi-x)\sin nx\,dx$$

$$=\left[x\left(-\frac{\cos nx}{n}\right)-1\left(-\frac{\sin nx}{n^2}\right)\right]_0^{\pi/2}$$

$=0$ at 0 $\quad =0$ at 0

$$+\left[(\pi-x)\left(-\frac{\cos nx}{n}\right)+1\left(\frac{\sin nx}{n^2}\right)\right]_{\pi/2}^{\pi}$$

$=0$ at π $\quad 0$ at π

$$=-\frac{\pi}{2n}\cos\frac{n\pi}{n}+\frac{1}{n^2}\sin\frac{n\pi}{2}+\frac{\pi}{2n}\cos\frac{n\pi}{2}+\frac{1}{n^2}\sin\frac{n\pi}{2}$$

$$=\frac{2}{n^2}\sin\frac{n\pi}{2}.$$

(b) The finite Fourier cosine transforms:

$$F_c(n)=\int_0^{\pi} f(x)\cos nx\,dx=\int_0^{\pi/2} x\cos nx\,dx$$

$$+\int_0^{\pi/2}(\pi-x)\cos nx\,dx$$

$$=\left[x\frac{\sin nx}{n}-1\left(-\frac{\cos nx}{n^2}\right)\right]_0^{\pi/2}$$

$$+\left[(\pi-x)\left(-\frac{\sin nx}{n}\right)-(-1)\left(-\frac{\cos nx}{n^2}\right)\right]_{\pi/2}^{\pi}$$

$$=\frac{\pi}{2n}\sin\frac{n\pi}{n}+\frac{1}{n^2}\left(\cos\frac{n\pi}{2}-1\right)+\frac{\pi}{2n}\sin\frac{n\pi}{2}$$

$$+\frac{1}{n^2}\left(\cos\frac{n\pi}{2}-(-1)^n\right)$$

$$=\frac{\pi}{n}\sin\frac{n\pi}{n}+\frac{1}{n^2}\left(2\cos\frac{n\pi}{2}-\left(1+(-1)^n\right)\right)$$

$$=\begin{cases}\frac{2}{n^2}\left((-1)^{n/2}-1\right),\text{if n is even}\\ \frac{\pi}{2}(-1)^{(n-1)/2},\text{if n is odd}\end{cases}$$

4

Partial Differential Equations

4.1 Formation of Partial Differential Equation

Here let us discuss to form a partial differential equation by eliminating arbitrary constants a and b from $z=(x+a)^2+(y+b)^2$.

$$p=\frac{\partial z}{\partial x}=2(x+a),q=\frac{\partial z}{\partial x}=2(y+b),4z=p^2+q^2$$

Problems

1. Let us discuss to form a partial differential equation by eliminating arbitrary constants a and b from $z = xf(2x+y)+g(2x+y)$.

Given: $Z = xf(2x+y)+g(2x+ y)$ ... 1 ...(1)

Solution:

Differentiate (1) partially with respect to x, y we get:

$$p=2xf'+2g'+f \quad \text{...(2)}$$

$$q=xf'+g' \quad \text{...(3)}$$

Differentiate (2) partially with respect to x:

$$r=4xf'' +2f'+4g''+ 2f'$$

$$r=4xf'' +4g''+4f \quad \text{... (4)}$$

Differentiate (2) partially with respect to y:

$$s=2xf'+2g''+f' \quad \text{...(5)}$$

Differentiate (3) partially with respect to y:

$$t=xf+g'' \quad \text{...(6)}$$

$$4\cdot(5)-(4)\Rightarrow A\,xf''+4g''=4s-r$$

$$4s - r = 4xf'' + 4g''$$

$$= 4[xf'' + g'']$$

$$= 4[t]$$

i.e., $4s - r - 4f = 0$ which is the required p.d.e.

2. Let us discuss to form the partial differential equation by eliminating arbitrary functions f and φ from $z = f(x+ct) + \varphi(x-ct)$.

Given: $z = f(x+ct) + \varphi(x-ct)$ ...(1)

Solution:

$$\frac{\partial z}{\partial x} = f'(x+ct) + \phi'(x-ct) \quad ...(2)$$

$$\frac{\partial z}{\partial t} = cf'(x+ct) - c\phi'(x-ct) \quad ...(3)$$

$$\frac{\partial^2 z}{\partial x^2} = f''(x+ct) - c\phi''(x-ct) \quad ...(4)$$

$$\frac{\partial^2 z}{\partial t^2} = c^2 f''(x+ct) + c^2\phi''(x-ct)$$

$$= c^2\left[f''(x+ct) + \phi''(x-ct)\right]$$

$$\frac{\partial^2 z}{\partial t^2} = c^2 \frac{\partial^2 z}{\partial x^2} \text{ by}(4)$$

Which is the required p.d.e.

3. Here let us discuss to form a partial differential equation by eliminating a and b from the expression $(x-a)^2 + (y-b)^2$.

Solution:

$$(x+a)^2 + (y-b)^2 + z^2 = c^2 \quad ...(1)$$

Diff equation (1) w.r.t 'x' partially:

$$2(x-a) + 2zp = 0$$

$$(x-a) = -zp \quad ...(2)$$

Diff equation (1) w.r.t 'y' partially:

$$2(y-a)+2zq=0$$

$$(y-a)= -zq \quad ...(3)$$

Substitute equation (2) and (3) in (1):

$$z^2p^2+z^2q^2+z^2=c^2$$

4.2 Solutions of First Order Linear (Lagrange) Equation and Non-linear Equations (Charpit's Method)

Solutions of First Order Linear (Lagrange) Equation

Problems

1. Let us now discuss to solve the Lagrange's linear equation of $x(y^2+z)p+y(x^2+z)q=z(x^2-y^2)$.

Solution:

$$\frac{dx}{x(y^2+z)}=\frac{dy}{y(x^2+z)}=\frac{dz}{z(x^2-y^2)}$$

Choose the multipliers $\left(\frac{1}{x},\frac{1}{y},\frac{1}{z}\right)$

$$\text{Each term}=\frac{\frac{1}{x}dx-\frac{1}{y}dy+\frac{1}{z}dz}{(y^2+z)-(x^2+z)+(x^2-y^2)}$$

$$\Rightarrow \int\frac{dx}{x}-\int\frac{dy}{y}+\int\frac{dz}{z}$$

$$\log\frac{xz}{y}=\log C_1$$

$$(\text{i.e})\frac{xy}{z}=\log C_1$$

Using the multipliers (x, -y, -1)

$$\text{Each term}=\frac{xdx-ydy-dz}{x^2y^2+x^2z-y^2x^2-y^2z-zx^2-zy^2}$$

$$\Rightarrow \int xdx-\int ydy-\int dz=0$$

$$x^2 - y^2 - 2z = C_2$$

Solution is $\phi\left(\frac{xy}{y}, x^2 - y^2 - 2z\right) = 0$

2. Let us see an example for linear equation of $\left(x^2 - y^2 - z^2\right)p + 2xyq = 2zx$.

Solution:

The given equation is of the form $P_p + Q_q = R$.

(Lagrangian form):

Here $P = \left(x^2 - y^2 - z^2\right)$; $Q = 2xy$ $R = 2zx$.

Subsidiary equation is $\frac{dx}{P} = \frac{dy}{Q} = \frac{dz}{R}$.

$$\text{i.e.,} \frac{dx}{\left(x^2 - y^2 - z^2\right)} = \frac{dy}{2xy} = \frac{dz}{2zx}$$

Grouping:

$$\frac{dy}{2xy} = \frac{dz}{2zx}$$

$$\Rightarrow \frac{dy}{y} - \frac{dz}{z} = 0$$

Integrating we get:

$$\log y - \log z = \log c_1$$

$$\log\left(\frac{y}{z}\right) = \log c_1$$

$$\Rightarrow u = y/z$$

Choose the multiplier x, y, z each and every ratio equal to:

$$\frac{x\,dx + y\,dy + z\,dz}{x^3 + xy^2 + xz^2}$$

Grouping:

$$\frac{xdx + ydy + zdz}{x\left(x^2 + y^2 + z^2\right)} = \frac{dy}{2xy}$$

$$\Rightarrow \frac{d\left(x^2+y^2+z^2\right)}{2x\left(x^2+y^2+z^2\right)}=\frac{dy}{2xy}$$

$$\therefore\ d\left(x^2+y^2+z^2\right)=2xdx+2ydy+2zdz$$

$$\Rightarrow \frac{d\left(x^2+y^2+z^2\right)}{2x\left(x^2+y^2+z^2\right)}-\frac{dy}{y}=0$$

Integrating:

$$\log\left(x^2+y^2+z^2\right)-\log y=\log C_2$$

$$\log\frac{\left(x^2+y^2+z^2\right)}{y}=\log C_2$$

$$\Rightarrow v=\frac{x^2+y^2+z^2}{y}$$

$\therefore$ Solution of equation (1) is f(u, v) = 0

$$\text{i.e.,} f\left(\frac{y}{z}\frac{x^2+y^2+z^2}{y}\right)=0$$

3. Let us discuss an example of Lagrange's equation of $(3z-4y)p+(4x-2z)q=(2y-3x)$.

Solution:

$$(3z-4y)p+(4x-2z)q=2y-3x \quad \ldots\ 1 \qquad \ldots(1)$$

This equation is of the form of $P_p+Q_q=R$.

Here $P=(3z-4y)$; $Q=(4x-2z)$ and $R=(2y-3x)$.

The subsidiary equation of (1) is:

$$\frac{dx}{P}=\frac{dy}{Q}=\frac{dz}{R}$$

$$\text{i.e.,}\frac{dx}{3z-4y}=\frac{dy}{4x-2z}=\frac{dz}{2y-3x} \qquad \ldots(2)$$

Case (i): Choose the Lagrangian multipliers (x, y, z) such that each and every ratio of (2) is equal to:

$$=\frac{x\,dx+y\,dy+z\,dz}{x(3z-4y)+y(4x-2z)+(y-3x)}=\frac{x\,dx+y\,dy+z\,dz}{0}$$

i.e., $x\,dx + y\,dy + z = 0$

Integrating we get:

$$\frac{x^2}{2}+\frac{y^2}{2}+\frac{z^2}{2}=C_1 \Rightarrow x^2+y^2+z^2=2C_1$$

$$\text{i.e.,} u=x^2+y^2+z^2$$

Case (ii): Choose another set of multipliers (2, 3, 4) such that each and every ratio of (2) is equal to:

$$=\frac{2dx+3dy+4dz}{2(3z-4y)+3(4x-2z)+4(2y-3x)}=\frac{2dx+3dy+4dz}{0}$$

i.e., $2\,dx+3\,dy+4\,dz=0$

Integrating we get:

$$2x + 3y + 4z = c$$

$$\text{i.e.,} v=2x+3y+4z$$

Hence the solution of the equation $P_p+Q_q=R$ is of the form f(u, v) = 0.

i.e., $f\left(x^2+y^2+z^2;\ 2x+3y+4z\right)=0$ is the required solution.

Here f is arbitrary function.

4. Let us solve an example of linear equation $x\left(z^2-y^2\right)p+y\left(x^2-z^2\right)q=z\left(y^2-x^2\right)$.

Solution:

$$x\left(z^2-y^2\right)p+y\left(x^2-z^2\right)q=z\left(y^2-x^2\right) \qquad \text{...(1)}$$

This equation is of the form $P_p + Qq = R$

Where $P=x\left(z^2-y^2\right);\ Q=y(x^2);\ R=z\left(y^2-x^2\right)$

Lagrange's subsidiary equations are:

$$\frac{dx}{P}=\frac{dy}{Q}=\frac{dz}{R}$$

$$\text{i.e.,}\frac{dx}{x\left(z^2-y^2\right)}=\frac{dy}{y\left(x^2-z^2\right)}=\frac{dz}{z\left(y^2-x^2\right)} \qquad ...(2)$$

Case (i): Choose the Lagrange multiplier (x, y, z) such that each and every ratio of (2) equal to:

$$\frac{\frac{1}{x}dx+\frac{1}{y}dy+\frac{1}{z}dz}{\left(z^2-y^2\right)+\left(x^2-z^2\right)+\left(y^2-x^2\right)}=\frac{\frac{1}{x}dx+\frac{1}{y}dy+\frac{1}{2}dz}{0}$$

$$\text{i.e., } u=x^2+y^2+z^2$$

Integrating we get:

$$\frac{x^2}{2}+\frac{y^2}{2}+\frac{z^2}{2}=C_1 \Rightarrow x^2+y^2+z^2=2C_1$$

$$\text{i.e.,} u=x^2+y^2+z^2$$

Case (ii): Choose another set of multiplier $\left(\frac{1}{x},1/y,1/z\right)$ such that each and every ratio of (2) equal to:

$$=\frac{x\,dx+y\,dy+z\,dz}{x^2\left(z^2-y^2\right)+y^2\left(x^2-z^2\right)+z^2\left(y^2-x^2\right)}=\frac{x\,dx+y\,dy+z\,dz}{0}$$

$$\text{i.e.,}\frac{1}{x}dx+\frac{1}{y}dy+\frac{1}{z}dz=0$$

Integrating we get:

$$\log x+\log y+\log z=\log C_2$$

$$\log xyz=\log C_2$$

$$\Rightarrow xyz=C_2$$

$$\text{i.e., } v=xyz$$

Solution of (1) is $\Phi(u,v)$:

$$\Phi\left(x^2+y^2+z^2,\ xyz\right)=0$$

5. Let us see an example of Lagrange's equation $(y-xz)p+(yz-x)q=(x+y)(x-y)$.

Solution:

$$(y-xz)p+(yz-x)q=(x+y)(x-y) \qquad \ldots\ 1 \qquad \ldots(1)$$

is of the form $P_p + Q_q = R$ (Lagrange's form)

Here $P=(y-xz)$; $Q=(yz-x)$ and $R=(x+y)(x-y)$

The subsidiary equation:

$$\frac{dx}{y-xz}=\frac{dy}{yz-x}=\frac{dz}{(x+y)(x-y)} \qquad \ldots(2)$$

Solutions of (1) is of the form $Q(u,\ v)=0$.

Case (i): Choose the multiplier (x, y, z) such that each and every ratio of (2) is equal to:

$$\frac{x\,dx+y\,dy+z\,dz}{xy-x^2z+y^2z-xy+x^2z-y^2z}=\frac{x\,dx+y\,dy+z\,dx}{0}$$

$$x\,dx+y\,dy+z\,dz=0$$

Integrating we get:

$$\frac{x^2}{2}+\frac{y^2}{2}+\frac{z^2}{2}\Rightarrow C_1$$

$$x^2+y^2+z^2=2C_1$$

$$\therefore\ u=x^2+y^2+z^2$$

Case (ii): Choose the multiplier (y, x, 1) such that each ratio of (2) equal to:

$$=\frac{y\,dx+x\,dy+dz}{y^2-xyz+xyz-x^2+x^2-y^2}=\frac{y\,dx+x\,dy+dz}{0}$$

$$y\,dx+x\,dy+dz=0 \quad \text{i.e., } d(xy)+dz=0$$

Integrating:

$$xy + z = C_2$$

$$\therefore V = xy + z$$

Solution of (1) is $Q\left(x2 + y2 + z2,\ xy + z\right) = o$.

Non-linear Equations (Charpit's Method)

We present here a general method for solving non-linear partial differential equations. This is known as Charpit's method.

Let,

$$F\left(x, y, u, p.q\right) = o$$

Be a general non-linear partial differential equation of first-order. Since u depends on x and y, we have:

$$du = u_x\, dx + u_y dy = p\, dx + q\, dy$$

Where $p = u_x = \frac{\partial u}{\partial x}$, $q = u_y = \frac{\partial u}{\partial y}$

If we can find another relation between x, y, u, p, q such that

$$f\left(x, y, u, p, q\right) = o$$

Then we can solve for p and q and substitute them in equation.

To determine f, differentiate w.r.t. x and y so that:

$$\frac{\partial F}{\partial x} + \frac{\partial F}{\partial u}p + \frac{\partial F}{\partial p}\frac{\partial p}{\partial x} + \frac{\partial F}{\partial q}\frac{\partial q}{\partial x} = o$$

$$\frac{\partial f}{\partial x} + \frac{\partial f}{\partial u}p + \frac{\partial f}{\partial p}\frac{\partial p}{\partial x} + \frac{\partial f}{\partial q}\frac{\partial q}{\partial x} = o$$

$$\frac{\partial F}{\partial y} + \frac{\partial F}{\partial u}q + \frac{\partial F}{\partial p}\frac{\partial p}{\partial y} + \frac{\partial F}{\partial q}\frac{\partial q}{\partial y} = o$$

$$\frac{\partial f}{\partial y} + \frac{\partial f}{\partial u}q + \frac{\partial f}{\partial p}\frac{\partial p}{\partial y} + \frac{\partial f}{\partial q}\frac{\partial q}{\partial y} = o$$

Eliminating $\frac{\partial p}{\partial x}$ from, equations and $\frac{\partial q}{\partial y}$ from equations we obtain:

$$\left(\frac{\partial F}{\partial x}\frac{\partial f}{\partial p}-\frac{\partial f}{\partial x}\frac{\partial F}{\partial p}\right)+\left(\frac{\partial F}{\partial u}\frac{\partial f}{\partial p}-\frac{\partial f}{\partial u}\frac{\partial F}{\partial p}\right)p+\left(\frac{\partial F}{\partial q}\frac{\partial f}{\partial p}-\frac{\partial f}{\partial q}\frac{\partial F}{\partial p}\right)\frac{\partial q}{dx}=o$$

$$\left(\frac{\partial F}{\partial y}\frac{\partial f}{\partial q}-\frac{\partial f}{\partial y}\frac{\partial F}{\partial q}\right)+\left(\frac{\partial F}{\partial u}\frac{\partial f}{\partial p}-\frac{\partial f}{\partial u}\frac{\partial F}{\partial p}\right)q+\left(\frac{\partial F}{\partial q}\frac{\partial f}{\partial q}-\frac{\partial f}{\partial q}\frac{\partial F}{\partial p}\right)\frac{\partial q}{dy}=o$$

Adding these two equations and using:

$$\frac{\partial q}{\partial x}=\frac{\partial^2 u}{\partial x\partial y}=\frac{\partial p}{\partial y}$$

and rearranging the terms, we get:

$$\left(-\frac{\partial F}{\partial p}\right)\frac{\partial F}{\partial x}+\left(-\frac{\partial F}{\partial q}\right)\frac{\partial f}{\partial q}+\left(-p\frac{\partial F}{\partial p}-q\frac{\partial F}{\partial q}\right)\frac{\partial f}{\partial u}+\left(\frac{\partial F}{\partial x}+p\frac{\partial F}{\partial u}\right)\frac{\partial f}{\partial p}+\left(\frac{\partial F}{\partial y}+q\frac{\partial f}{\partial u}\right)\frac{\partial f}{\partial q}=o$$

Following arguments we get the auxiliary system of equations:

$$\frac{dx}{\frac{-\partial F}{\partial p}}=\frac{dy}{\frac{-\partial F}{\partial q}}=\frac{du}{-p\frac{\partial F}{\partial p}-q\frac{\partial F}{\partial q}}=\frac{dp}{\frac{\partial F}{\partial x}+p\frac{\partial F}{\partial u}}=\frac{dq}{\frac{\partial F}{\partial y}+q\frac{\partial F}{\partial u}}=\frac{df}{o}$$

An Integral of these equations, involving p or q or both, can be taken as the required equation.

Problem

1. Find the general solution of the partial differential equation.

$$\left(\frac{\partial u}{\partial x}\right)^2 x+\left(\frac{\partial u}{\partial y}\right)^2 y-u=o$$

Solution:

Let $p=\frac{\partial u}{\partial x},\ q=\frac{\partial u}{\partial y}$

The auxiliary system of equations is:

$$\frac{dx}{2px}=\frac{dy}{2qy}=\frac{du}{2\left(p^2x+q^2y\right)}=\frac{dp}{p-p^2}=\frac{dq}{q-q^2}$$

Which we obtain by putting values of:

$$\frac{\partial F}{\partial p} = 2px, \frac{\partial F}{\partial q} = 2qy, \frac{\partial F}{\partial x} = p^2, \frac{\partial F}{\partial x} = -1, \frac{\partial f}{\partial y} = q^2$$

And multiplying by -1 throughout the auxiliary system. From first and 4th expression we get:

$$dx = \frac{p^2dx + 2pxdp}{py}$$

From second and 5th expression:

$$dy = \frac{q^2dy + 2qydq}{qy}$$

Using these values of dx and dy we get:

$$\frac{p^2dx + 2pxdp}{p^2x} = \frac{p^2dy + 2qydq}{q^2y}$$

Or,

$$\frac{dx}{x} + \frac{2}{p}dp = \frac{dy}{y} + \frac{2dq}{q}$$

Taking integral of all terms we get:

$$\ln|x| + 2\ln|p| = \ln|y| + 2\ln|q| + \ln c$$

Or,

$$\ln|x|p^2 = \ln|y|q^2c$$

Or,

$p^2x = cq^2y$, where c is an arbitrary constant.

Solving for p and q we get $cq^2y + q^2y - u = 0$

$$(c+1)q^2y = u$$

$$q = \left\{\frac{u}{(c+1)y}\right\}^{1/2}$$

$$p = \left\{ \frac{u}{(c+1)x} \right\}^{1/2}$$

Takes the following form in this case:

$$du = \left\{ \frac{u}{(c+1)x} \right\}^{1/2} dx + \left\{ \frac{u}{(c+1)y} \right\}^{1/2} dy$$

Or,

$$\left(\frac{1+c}{u} \right)^{1/2} du = \left(\frac{c}{x} \right)^{1/2} dx + \left(\frac{i}{y} \right)^{1/2} dy$$

By integrating this equation we obtain $((1+c)u)^{1/2} = (cx)^{1/2} = (y)^{1/2} + C_1$

This is a complete solution.

4.3 Method of Separation of Variables for Second Order Equations

If one can re-arrange the ordinary differential equation into the following standard form $\frac{dy}{dx} = f(x)g(y)$, then the solution may be found by the technique of separation of variables.

$$\int \frac{dy}{g(y)} = \int f(x)dx$$

This result is obtained by dividing the standard form by g(y) and then integrating both sides with respect to x.

This method consists of the following steps:

- If x and y are independent variables and z is the dependent variable, we find a solution of the given equation in the form z=XY, where X=X(x) is a function of x alone and Y=Y(y) is a function of y alone. Then, we substitute for z and its partial derivative (computed from z=XY) in the equation and rewrite the equation in such a way that the L.H.S involves X and its derivatives and the R.H.S involves Y and its derivatives.
- We equate each side of the equation obtained in step 1 to a constant and solve the resulting O.D.E for X and Y.

- Finally we substitute the expression for X and Y obtained in step 2 in z=XY. The resulting expression is the general solution for z.

Problems

1. Let us discuss the partial differential equation by eliminating the arbitrary constants a and b from $z=(x^2+a)(y^2+b)$.

Given: $Z=(x^2+a)(y^2+b)$ ...(1)

Solution:

Differentiate (1) point w.r.t. x and y:

$$p=2x(y^2+b)\Rightarrow(y^2+b)=\frac{p}{2x} \quad \text{...(2)}$$

$$q=2y(x^2+a)\Rightarrow(x^2+a)=\frac{q}{2y} \quad \text{...(3)}$$

Substitute (2) & (3) in (1)

$$Z=\left(\frac{q}{2v}\right)\left(\frac{p}{2x}\right) \text{ i.e., } 4xyz=pq$$

2. By the method of separation of variables solve the question $\frac{\partial z}{\partial x}y^3+\frac{\partial z}{\partial y}x^3=0$

Solution:

Let the solution be in the form z=XY where X=X(x) and Y=Y(y)

$$\Rightarrow\frac{\partial z}{\partial x}=\frac{\partial X}{\partial x}Y \text{ and } \frac{\partial z}{\partial x}=X\frac{\partial Y}{\partial y}$$

Puting these in the given equation we get:

$$\left(\frac{\partial X}{\partial x}\right)y^3+\left(X\frac{\partial Y}{\partial y}\right)x^2=0 \text{ or}$$

$$\left(\frac{dX}{dx}Y\right)y^3=-\left(X\frac{dY}{dy}\right)x^2$$

$$\frac{1}{X}\left(\frac{dX}{dx}\right)\frac{1}{x^2}=-\frac{1}{Y}\left(\frac{dY}{dy}\right)\frac{1}{y^3}$$

L.S.H is a function of x only and R.S.H is a function of y only.

Since x and y are independent variable, this expression can hold only if each side is a constant i.e.,

$$\frac{1}{x}\left(\frac{dX}{x}\right)\frac{1}{x^2}=k \text{ and} -\frac{1}{y}\left(\frac{dy}{dy}\right)\frac{1}{y^3}= k \text{ (Where k is constant)}$$

These may be written as:

$$\frac{1}{x}\left(\frac{dX}{dx}\right)=kx^2 \text{ and } \frac{1}{y}\left(\frac{dY}{dy}\right)=-ky^3$$

$$\frac{d}{dx}(\log X)=kx^2 \text{ and } \frac{d}{dy}(\log Y)=-ky^3$$

Integrating w.r.t x and y we get:

$$\log X = k\frac{x^3}{3}+\log C_1 \text{ and } \log Y = -k\frac{y^4}{4}+\log C_2$$

$$\Rightarrow X = C_1 e^{k\frac{x^3}{3}} \text{ and } Y = C_2 e^{-k\frac{y^4}{4}}$$

Where C_1 and C_2 are constants:

$$\therefore z = XY = Ae^{k\frac{x^2}{3}-k\frac{y^4}{4}} \text{ where } A = C_1C_2$$

3. Using the method of separation of variable solve $\frac{\partial u}{\partial x}=2\frac{\partial u}{\partial t}+u$ where $(x,o)=6e^{-3x}$

Solution:

Here x and t are independent variables and u id the dependent variable:

Let the solution be in the form $u = X(x)Y(t)$

$$\therefore \frac{\partial u}{\partial x}=\frac{dX}{dx}Y(t) \text{ and } \frac{\partial u}{\partial t}=X(x)\frac{dY}{dt}$$

The given equation takes the form:

$$\frac{dX}{dx}y(t)=2X(x)\frac{dy}{dt}+X(x)Y(t)$$

$$\left(\frac{dX}{dx}-X\right)Y=2X\frac{dY}{dt}$$

$$\frac{1}{X}\frac{dX}{dx}-1=\frac{2}{Y}\frac{dY}{dt}$$

$$\Rightarrow \frac{1}{X}\frac{dX}{dx}-1=k \text{ and } \frac{2}{Y}\frac{dY}{dt}=k$$

$$\frac{dX}{X}=(k+1)dx \text{ and } \frac{dY}{Y}\ \frac{k}{2}dt$$

On integration we get:

$$\log X=(k+1)x+\log C_1 \text{ and } \log Y=\frac{k}{2}t+\log C_2$$

$$\Rightarrow X=C_1e^{(k+1)x} \text{ and } Y=C_2e^{\frac{k}{2}t}$$

$\therefore$ The required solution is:

$$u=XY=Ae^{(k+1)x}\,e^{\frac{k}{2}t} \text{ Where } A=C_1C_2$$

But $\quad u(x,0)=6e^{-3x}$

$$\Rightarrow 6e^{-3x}=Ae^{(k+1)x}$$

$$\Rightarrow A=6 \text{ and } k+1=-3$$

$$\Rightarrow A=6 \text{ and } k=-4$$

$$\therefore u=6e^{-(3x+2t)}$$

4. Solve $\frac{\partial u}{\partial x}=\frac{\partial u}{\partial t}+u$

Solution:

Let the solution be in the form u = XT where X=X(x) and T=T(t):

$$\frac{\partial u}{\partial x}=T\frac{\partial X}{\partial x} \text{ and } \frac{\partial u}{\partial t}=X\frac{\partial T}{\partial t}$$

Putting these in the given equation we get:

$$T\frac{\partial X}{\partial x}=2X\frac{\partial T}{\partial t}+XT \text{ or } \frac{1}{X}\frac{\partial X}{\partial x}=\frac{2}{T}\frac{\partial T}{\partial t}+1$$

LHS is a function of X only and RHS is of T alone:

Since x and t are independent variables, this expression can hold only if each side is a constant i.e:

$$\frac{1}{X}\frac{dX}{dx} = k \text{ and } \frac{2}{T}\frac{dT}{dt} + 1 = k$$

$$\frac{1}{X}dX = kdx \text{ and } \frac{1}{T}dT = \frac{k-1}{2}dt$$

$$\operatorname{Log} X = kx + \operatorname{Log} c_1 \text{ and } \operatorname{Log} T = \frac{k-1}{2}t + \operatorname{Log} c_2$$

$$\operatorname{Log}\frac{X}{c_1} = kx \text{ and } \operatorname{Log}\frac{T}{c_2} = \frac{k-1}{2}t$$

$$X = c_1 e^{kx} \text{ and } T = c_2 e^{\frac{k-1}{2}t}$$

Then the complete solution is given by:

$$u = XT = c_1 e^{kx} c_2 e^{\frac{k-1}{2}} = ce^{kx - \frac{k-1}{2}t}$$

5. Solve xp = yq by the method of separation of variables.

Solution:

We have the PDE: $x\frac{\partial z}{\partial x} = y\frac{\partial z}{\partial y}$

Let the solution be in the form z=XY where X = X(x) and Y = Y(y):

$$\frac{\partial u}{\partial x} = Y\frac{\partial X}{\partial x} \text{ and } \frac{\partial u}{\partial y} = X\frac{\partial Y}{\partial y}$$

Putting these in the given equation we get:

$$xY\frac{\partial X}{\partial x} = yX\frac{\partial Y}{\partial y} \Rightarrow \frac{x}{X}\frac{\partial X}{\partial x} = \frac{y}{Y}\frac{\partial Y}{\partial y}$$

$$\frac{x}{X}\frac{dX}{dx} = k \text{ and } \frac{y}{Y}\frac{dY}{dy} = k$$

$$\frac{dX}{X} = k\frac{dx}{x} \text{ and } \frac{dY}{Y} = k\frac{dy}{y}$$

$$\mathrm{Log}X = k\mathrm{Log}x + \mathrm{Log}c_1 \text{ and } \mathrm{Log}Y = k\mathrm{Log}y + \mathrm{Log}c_2$$

$$X = x^k c_1 \text{ and } Y = y^k c_2$$

The solution is given by:

$$z = XY = cx^k y^k = c(xy)^k$$

6. Solve $\frac{\partial^2 z}{\partial x \partial y} = z$

Solution:

Let the solution be in the form z = XY where X=X(x) and Y=Y(y):

$$\frac{\partial z}{\partial y} = X\frac{\partial Y}{\partial y} \text{ and } \frac{\partial}{\partial x}\left(\frac{\partial z}{\partial y}\right) = \frac{\partial^2 z}{\partial x \partial y} = \frac{\partial X}{\partial x}\frac{\partial Y}{\partial y}$$

Putting these in the given equation we get:

$$\frac{\partial X}{\partial x}\frac{\partial Y}{\partial y} = XY \Rightarrow \frac{1}{x}\frac{\partial X}{\partial x} = \frac{Y}{\frac{\partial Y}{\partial y}}$$

LHS is a function of X only and RHS is of Y alone.

Since x and y are independent variables, this expression only,

If each side is a constant i.e:

$$\frac{1}{X}\frac{dX}{dx} = k \text{ and } \frac{Y}{\frac{dY}{dy}} = k$$

$$\frac{1}{X}dX = kdx \text{ and } \frac{1}{k}dy = \frac{dY}{Y}$$

$$\mathrm{Log}X = kx + \mathrm{Log}c_1 \text{ and } \mathrm{Log}Y = \frac{Y}{K} + \mathrm{Log}C_2$$

$$X = c_1 e^{kx} \text{ and } Y = c_2 e^{\frac{y}{k}}$$

∴ The solution is given by:

$$Z = XY = c_1 e^{kx}.c_2 e^{\frac{y}{k}} = ce^{kx - \frac{y}{k}}$$

7. Solve $\frac{\partial^2 z}{\partial x^2}-2\frac{\partial z}{\partial x}+\frac{\partial z}{\partial y}=0$

Solution:

Let the solution be in the form z = XY where X=X(x) and Y=Y(y):

$$\frac{\partial z}{\partial x}=Y\frac{\partial X}{\partial x},\frac{\partial^2 z}{\partial x^2}=Y\frac{\partial^2 X}{\partial x^2} \text{ and } \frac{\partial z}{\partial y}=X\frac{\partial Y}{\partial y}$$

Substituting these in the given equation we get:

$$Y\frac{\partial^2 X}{\partial x^2}-2Y\frac{\partial X}{\partial x}+X\frac{\partial Y}{\partial y}=0$$

$$Y\left(\frac{\partial^2 x}{\partial x^2}-2\frac{\partial X}{\partial x}\right)=-X\frac{\partial Y}{\partial y}$$

$$\Rightarrow\left(\frac{\partial^2 x}{\partial x^2}-2\frac{\partial X}{\partial x}\right)=-\frac{1}{y}\frac{\partial Y}{\partial y}$$

LHS is a function of x only RHS is of Y alone.

Since x and y are independent variables, this expression can holds only if each side is a constant i.e.,

Consider $\frac{1}{x}\left(\frac{d^2X}{dx^2}-2\frac{dX}{dx}\right)=k$

$$\frac{d^2X}{dx^2}-2\frac{dX}{dx}-Xk-0$$

$$m^2-2m-k=0$$

Roots are m $=1\pm\sqrt{1+k}$

$$\therefore X=c_1e^{\left(1+\sqrt{1+k}\right)x}+c_2e^{\left(1-\sqrt{1+k}\right)x}$$

Considering $-\frac{1}{y}\frac{dY}{dy}=k \Rightarrow \frac{1}{Y}dY=-kdy$

$$\text{Log}Y=-ky+\text{Log}c_2$$

$$Y=c_3e^{-ky}$$

The complete solution is given by:

$$Z = XY = \left(c_1 e^{(1+\sqrt{1+k})} + c_2 e^{(1-\sqrt{1+k})x}\right) c_3 e^{-ky}$$

4.4 Applications of Partial Differential Equations

Consider a thin elastic membrane stretched tightly over a rectangular frame. Suppose the dimensions of the frame are a × b and that we keep the edges of the membrane fixed to the frame.

Perturbing the membrane from equilibrium results in some sort of vibration of the surface.

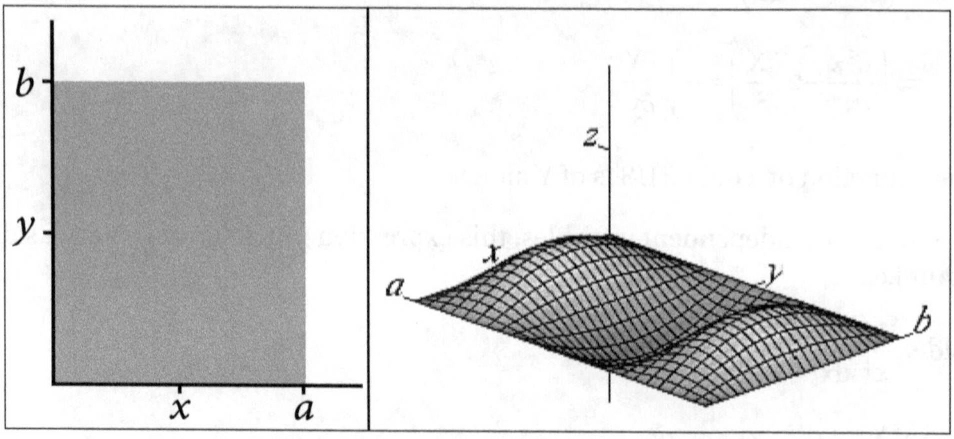

Our goal is to mathematically model the vibrations of the membrane surface.

We let:

$u(x,y,t)$ = deflection of membrane from equilibrium at position (x,y) and time t.

For a fixed t, the surface z = u(x, y, t) gives the shape of the membrane at time t.

Under ideal assumptions, one can show that u satisfies the two dimensional wave equations:

$$U_{tt} = c^2 \nabla^2 u = c^2 \left(U_{xx} + U_{yy}\right)$$

For $o < x < a, < o < y < b$

The fact that we are keeping the edges of the membrane fixed is expressed by the boundary conditions:

$$u(o,y,t)=u(a,y,t)=o \quad o\le y\le b,t\ge o$$
$$u(x,o,t)=u(x,b,t)=o \quad o\le x\le a,t\ge o$$

We must also specify how the membrane is initially deformed and set into motion. This is done via the initial conditions:

$$u(x,y,o)=f(x,y), \quad (x,y)\in R,$$
$$u_t(x,y,o)=g(x,y), \quad (x,y)\in R,$$

Where $R=[o,a]\times[o,b]$

4.4.1 Heat Equation

A partial differential diffusion equation of the form:

$$\frac{\partial U}{\partial t}=k\nabla^2 U \qquad ...(1)$$

Physically, the equation commonly arise in situations where k the normal diffusivity and U the temperature. The one dimensional heat conduction equation is:

$$\frac{\partial U}{\partial t}=k\frac{\partial^2 U}{\partial x^2} \qquad ...(2)$$

This can be solved by separation of variable using:

$$U(x,t)=X(x)T(t) \qquad ...(3)$$

Then:

$$X\frac{dT}{dt}=kT\frac{d^2X}{dx^2} \qquad ...(4)$$

Dividing both sides by kXT gives:

$$\frac{1}{kT}\frac{dT}{dt}=\frac{1}{X}\frac{d^2X}{dx^2}=-\frac{1}{\lambda^2} \qquad ...(5)$$

Where each side must be equal to a constant. Anticipating the exponential solution T, we have picked a negative separations constant so that the solution remains finite all times and λ has units of length.

The T solution is:

$$T(t)=Ae^{-xt/\lambda^2} \quad ...(6)$$

And the X solution is:

$$X(x)=B\cos\left(\frac{x}{\lambda}\right)+C\sin\left(\frac{x}{\lambda}\right) \quad ...(7)$$

The general solution is then:

$$U(x,t)=T(t)X(x) \quad ...(8)$$

$$=Ae^{-xt/\lambda^2}\left[B\cos\left(\frac{x}{\lambda}\right)+C\sin\left(\frac{x}{\lambda}\right)\right] \quad ...(9)$$

$$=e^{-xt/\lambda^2}\left[D\cos\left(\frac{x}{\lambda}\right)+E\sin\left(\frac{x}{\lambda}\right)\right] \quad ...(10)$$

If we are given the boundary condition:

$$U(o,t)=o \quad ...(11)$$

and,

$$U(L,t)=o \quad ...(12)$$

Then applying (11) to (10) gives:

$$D\cos\left(\frac{x}{\lambda}\right)=o\Rightarrow D=o \quad ...(13)$$

and applying (12) to (10) gives:

$$E\sin\left(\frac{L}{\lambda}\right)=o\Rightarrow n\pi\Rightarrow\lambda=\frac{L}{n\pi} \quad ...(14)$$

So (10) becomes:

$$U_\lambda(x,t)=E_\lambda e^{-k(x,\pi,L)^2}\sin\left(\frac{n\pi x}{L}\right) \quad ...(15)$$

Since the general solution can have any n:

$$U(x,t)=\sum_{n=1}^{\infty}c_s\sin\left(\frac{n\pi}{L}\right)e^{-k(n,\pi,L)^2t} \quad ...(16)$$

Now, if we are given an initial condition U(x,o) we have:

$$U(x,o)=\sum_{n=1}^{\infty}C_n\sin\left(\frac{n\pi x}{L}\right) \quad ...(17)$$

Multiplying both sides by $\sin(m\pi x/L)$ and integrating from O to L gives:

$$\int_o^L \sin\left(\frac{m\pi x}{L}\right)U(x,o)dx = \int_o^L \sum_{n=1}^{\infty} c_n \sin\left(\frac{m\pi x}{L}\right)\sin\left(\frac{n\pi x}{L}\right)dx \quad ...(18)$$

Using the orthogonally of sin (n,x) and sin (m,x):

$$\sum_{n=1}^{\infty} c_n \sin\left(\frac{n\pi x}{L}\right)\sin\left(\frac{m\pi x}{L}\right)dx = \sum_{n=1}^{\infty}\frac{1}{2}L\delta_{mn}c_n \quad ...(19)$$

$$=\frac{L}{2}c_m \quad ...(20)$$

$$=\int_o^L \sin\left(\frac{m\pi x}{L}\right)U(x,o)dx \quad ...(21)$$

So,

$$c_m = \frac{2}{L}\int_o^x \sin\left(\frac{m\pi x}{L}\right)U(x,o)dx \quad ...(22)$$

If the boundary condition are replaced by the requirements that the derivative of the temperature be zero at the edges then $(\odot)$ and $(\odot)$ are replaced by:

$$\frac{\partial U}{\partial x}\bigg|_{(o.t)} = 0 \quad ...(23)$$

$$\frac{\partial U}{\partial x}\bigg|_{(L.t)} = 0 \quad ...(24)$$

Following the same procedure as before, as smile answer is found, but with sin e replaced by cosine:

$$U(x,t) = \frac{1}{2}c_o + \sum_{n=1}^{\infty} c_s \cos\left(\frac{n\pi x}{L}\right)e^{-k(n,\pi,L)^2 t} \quad ...(25)$$

Where,

$$C_n = \frac{2}{L}\int_o^L \cos\left(\frac{m\pi x}{L}\right)U(x,o)dx \quad ...(26)$$

Problems

1. Let us discuss the ends A and B of a rod f c.m long have their temperature kept at 30°C and 80°C, until steady state conditions prevail. The temperature of the end B is suddenly reduced to 60°C and that of A is increased to 40°C. Find the temperature distribution in the rod after time t.

Solution:

Let u(x, t) be the temperature distribution at any distance 'x' at any time't' then u(x, t) satisfies one dimensional heat equation:

$$\frac{\partial u}{\partial t} = \alpha^2 \frac{\partial^2 u}{\partial x^2}$$

In steady state, the solution is:

$$u(x) = \frac{50x}{l} + 30$$

- $u(0,t) = 40$, for all $t \geq 0$
- $u(l,t) = 60$, for all $t \geq 0$
- $u(x,0) = \frac{50x}{\ell} + 30, \quad 0 \leq x \leq \ell$

Let $u(x,t) = u_s(x) + u_t(x,t)$

In steady state:

$$u_s(x) = \frac{20x}{\ell} + 40$$

In transient state, the conditions are:

- $u_,(0,t) = 0$
- $u_,(f,t) = 0$
- $u_t(x,0) = \frac{30}{\ell} - 10$

The most suitable solution is:

$$u_t(x,t) = e^{-\alpha^2 p^2 t}(A\cos px + B\sin px) \quad ...(1)$$

Substitute the first two boundary condition in equation 1.

$$u_t(x,t) = \sum_{n=1}^{\infty} b_n e^{-\alpha^2 p^2 t} \sin px$$

Where,

$$p = \frac{n\pi}{\ell}$$

Substitute the third boundary condition in the above equation.

$$b_n = \frac{2}{\pi}\int_0^{\pi} u_t(x,0)\sin\frac{n\pi}{\ell}xdx$$

$$= \frac{2}{\pi}\int_0^{\pi}\left(\frac{30x}{\ell} - 10\right)\sin\frac{n\pi}{\ell}xdx$$

$$b_n = -\frac{20}{n\pi}\left[1+2(-1)^n\right]$$

Hence the temperature is:

$$u(x,t) = \frac{20x}{\ell} + 40 - \frac{20}{\pi}\sum_{n=1}^{\infty}\frac{1}{n}\left(1+2(-1)^n\right)\sin\frac{n\pi}{\ell}xe^{-\frac{n^2\pi^2\alpha^2}{\iota}t}$$

2. A rod, 30 cm. Long, has its ends A and B kept at 20°C and 80°C respectively, until steady state conditions prevail. The temperature at each end is then suddenly reduced to 0°C and kept so. Find the resulting temperature function U(x, t) taking x = 0 at A.

Solution:

The temperature function u(x, t) is the solution of the one dimensional heat equation:

$$\frac{\partial u}{\partial t} = \alpha^2 \frac{\partial^2 u}{\partial x^2}$$

When the steady state condition prevails $\frac{\partial u}{\partial t} = 0$ and hence we get $\frac{\partial^2 u}{\partial x^2} = 0$ solving

$$u(x) = ax + b$$

When, $x = 0$; $u = 20 => u(0) = b = 20b = 20b = 20$

When, $x = l$; $u = 80 \Rightarrow u(l) = al + 20 = 80$

$$\Rightarrow \frac{60}{\iota} = a$$

Thus, $u(x,0) = \frac{60x}{l} + 20$

Boundary conditions are:

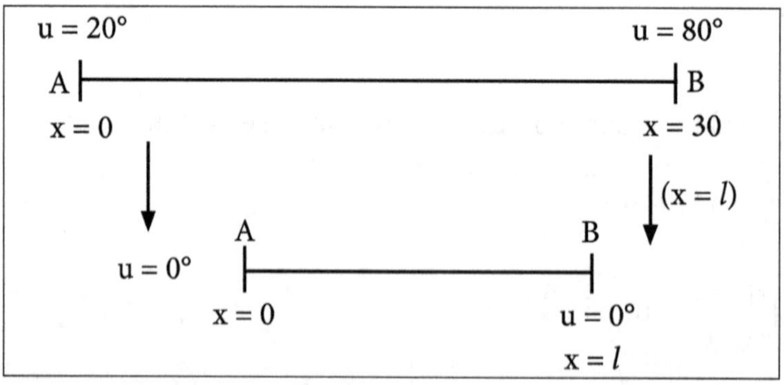

Let us consider length = 30 = l

(i) $u(0,t) = 0;\ t \geq 0$

(ii) $u(l, t) = 0;\ t > 0$

(iii) $u(x,0) = \frac{60x}{l} + 20\ ;0 \leq x \leq l$

Now the suitable solution which satisfies our boundary conditions is given by:

$$u(x,t) = (A\cos px + B\sin px)e^{-\alpha^2p^2t} \quad ...(1)$$

Applying condition (i) in (1), put x = 0

$$u(0,t) = Ae^{-\alpha^2p^2t} = 0$$

$\Rightarrow A = 0$ Sub in (1)

$$u(x,t) = B\sin px\ e^{-\alpha^2p^2t} \quad ...(2)$$

Applying (ii) in (2) i.e., put x = l

$$u(l,t) = B\sin pl\ e^{-\alpha^2p^2t} = 0$$

$$\Rightarrow \sin pl = 0$$

$$\Rightarrow p = \frac{n\pi}{l} \quad \text{i.e.,} p = \frac{n\pi}{l} \text{ sub in } (2).$$

$$u(x,t) = B\sin\frac{n\pi x}{l} e^{-\frac{\alpha^2n^2\pi^2t}{l^2}} \quad ...(3)$$

The most general solution is:

$$u(x,t)=\sum_{n=1}^{\infty}B_n \sin\frac{n\pi x}{l}e^{-\frac{\alpha^2 n^2\pi^2 t}{l^2}} \qquad ...(A)$$

Applying (iii) put t = 0 in (A)

$$f_{say}^{(x)}=u(x,0)=\sum_{n=1}^{\infty}B_n \sin\frac{n\pi x}{l} \qquad ...(4)$$

We know that half range Fourier sine series for f(x) in (0, 1) is:

$$f(x)=\sum_{n=1}^{\infty}b_n \sin\frac{n\pi x}{l} \qquad ...(5)$$

Where $b_n=\frac{2}{l}\int_0^1 f(x).\sin\frac{n\pi x}{l}dx$ ...(6)

From (4) & (5) we observe that:

$$B_n=b_n=\frac{2}{l}\int_0^1 f(x)\sin\frac{n\pi x}{l}dx$$

$$=\frac{2}{l}\int_0^1\left(\frac{60x}{1}+20\right)\sin\frac{n\pi x}{1}dx$$

$$=\frac{2}{l}\left[\left(\frac{60x}{l}+20\right)\left(\frac{-l}{n\pi}\cos\frac{n\pi x}{l}\right)-\left(\frac{60}{1}\right)\left(\frac{-1^2}{n^2\pi^2}\sin\frac{n\pi x}{l}\right)\right]_0^1$$

$$=\frac{2}{l}\left[\left(\frac{-80l}{n\pi}(-1)^n\right)-\left(\frac{-20\,l}{nip}\right)\right]$$

$$=\left(\frac{2}{1}\right)\left(\frac{201}{n\pi}\right)\left[1-4(-1)^n\right]$$

$$B_n=\frac{40}{n\pi}\left[1-4(-1)^n\right] \text{ sub in (A)}$$

$$u(x,t)=\sum_{n=1}^{\infty}\frac{40}{n\pi}\left(1-4(-1)^n\right)\sin\frac{n\pi x}{1}e^{-\frac{\alpha^2 n^2\pi^2 t}{1^2}}$$

Replace l by 30 cm:

$$u(x,t)=\frac{40}{\pi}\sum_{n=1}^{\infty}\frac{1}{n}\left[1-4(-1)^n\right]\sin\frac{n\pi x}{30}e^{-\frac{\alpha^2 n^2\pi^2 t}{900}}.$$

5

Vector Calculus

5.1 Scalar Point Function and Vector Point Function

Scalar Point Function

A scalar point function is a function which assigns a real number to each point of some region of the space. If each point (x, y, z) of a region R in space is assigned a real number $u = \phi\,(x, y, z)$, then ϕ is called as scalar point function.

Vector Point Function

A vector point function is the function which assigns a vector to each point of some region of the space. If each point (x, y, z) of a region R is in space, there is assigned a vector $F = F(x, y, z)$, then F is called as vector point function. Such a function would have a representation:

$$F = f_1(x, y, z)\,i + f_2(x, y, z)\,j + f_3(x, y, z)\,k$$

Or equivalently:

$$F = \begin{bmatrix} f_1 \\ f_2 \\ f_3 \end{bmatrix} = \begin{bmatrix} f_1(x,y,z) \\ f_1(x,y,z) \\ f_1(x,y,z) \end{bmatrix}$$

Syn. vector function of position.

5.1.1 Gradient, Divergence and Curl

Gradient

The Vector Differential Operator ∇ (read as del) is defined as $\nabla \equiv \vec{i}\frac{\partial}{\partial x} + \vec{j}\frac{\partial}{\partial y} + \vec{k}\frac{\partial}{\partial z}$

where $\left(\vec{i}, \vec{j}, \vec{k}\right)$ are unit vectors along the three rectangular axes OX, OY, OZ.

Properties of Gradient

Problems

1. If f and g are two scalar point function that $\nabla(f \pm g) = \nabla f \pm \nabla f \pm \nabla g$ or $\text{grad}(f \pm g)$ $= \text{fgradg} + \text{ggradf}$.

Solution:

$$\nabla(f \pm g) = \left(\vec{i}\frac{\partial}{\partial x} + \vec{j}\frac{\partial}{\partial y} + \vec{k}\frac{\partial}{\partial z}\right)(f \pm g)$$

$$= \left(\vec{i}\frac{\partial}{\partial x}(f \pm g) + \vec{j}\frac{\partial}{\partial y}(f \pm g) + \vec{k}\frac{\partial}{\partial z}(f \pm g)\right)$$

$$= \left(\vec{i}\frac{\partial}{\partial x}(f \pm g) + \vec{j}\frac{\partial}{\partial y}(f \pm g) + \vec{k}\frac{\partial}{\partial z}(f \pm g)\right)$$

$$= \vec{i}\frac{\partial f}{\partial x} \pm \vec{i}\frac{\partial g}{\partial x} + \vec{j}\frac{\partial f}{\partial y} \pm \vec{j}\frac{\partial g}{\partial y} + \vec{k}\frac{\partial f}{\partial z} \pm \vec{k}\frac{\partial g}{\partial z}$$

$$= \left(\vec{i}\frac{\partial f}{\partial x} + \vec{j}\frac{\partial f}{\partial y} + \vec{k}\frac{\partial f}{\partial z}\right) \pm \left(\vec{i}\frac{\partial g}{\partial x} + \vec{j}\frac{\partial g}{\partial y} + \vec{k}\frac{\partial g}{\partial z}\right)$$

$$= \nabla f \pm \nabla g$$

2. If f and g are two scalar point functions then $\nabla(fg) = f\nabla g + g\nabla f$ or $\text{grad}(fg) = \text{fgradg}$ $+ \text{ggradf}$.

Solution:

$$\nabla(fg) = \left(\vec{i}\frac{\partial}{\partial x} + \vec{j}\frac{\partial}{\partial y} + \vec{k}\frac{\partial}{\partial z}\right)(fg)$$

$$= \left(\vec{i}\frac{\partial}{\partial x}(fg) + \vec{j}\frac{\partial}{\partial y}(fg) + \vec{k}\frac{\partial}{\partial z}(fg)\right)$$

$$= \vec{i}\left(f\frac{\partial g}{\partial x} + g\frac{\partial f}{\partial x} + \right) + \vec{j}\left(f\frac{\partial g}{\partial y} + g\frac{\partial f}{\partial y}\right) + \vec{k}\left(f\frac{\partial f}{\partial z} + g\frac{\partial f}{\partial z}\right)$$

$$= f\left(\vec{i}\frac{\partial g}{\partial x} + \vec{j}\frac{\partial g}{\partial y} + \vec{k}\frac{\partial g}{\partial z}\right) + g\left(\vec{i}\frac{\partial f}{\partial x} + \vec{j}\frac{\partial f}{\partial y} + \vec{k}\frac{\partial f}{\partial z}\right)$$

$$= f\nabla g + g\nabla f$$

3. If f and g are two scalar points function then $\nabla\left(\frac{f}{g}\right) = \frac{g\nabla f - f\nabla g}{g^2}$ where $g \neq 0$.

Solution:

$$\nabla\left(\frac{f}{g}\right) = \left(\vec{i}\frac{\partial}{\partial x} + \vec{j}\frac{\partial}{\partial y} + \vec{k}\frac{\partial}{\partial z}\right)\left(\frac{f}{g}\right)$$

$$= \sum \vec{i}\frac{\partial}{\partial x}\left(\frac{f}{g}\right)$$

$$= \sum \vec{i}\left(\frac{g\frac{\partial f}{\partial x} - f\frac{\partial g}{\partial x}}{g^2}\right)$$

$$= \frac{1}{g^2}\left(g\sum \vec{i}\frac{\partial f}{\partial x} - f\sum \vec{i}\frac{\partial g}{\partial x}\right)$$

$$= \frac{1}{g^2}[g\nabla f - f\nabla g]$$

4. If $(\vec{r}) = (x\vec{i} + y\vec{j} + z\vec{k})$ such that $|\vec{r}| = r$, let us prove that $\nabla r^n = nr^{n-2}\vec{r}$.

Solution:

$$\nabla r^n = \left(\vec{i}\frac{\partial}{\partial x} + \vec{j}\frac{\partial}{\partial y} + \vec{k}\frac{\partial}{\partial z}\right)r^n$$

$$= \left(\vec{i}\frac{\partial r^n}{\partial x} + \vec{j}\frac{\partial r^n}{\partial y} + \vec{k}\frac{\partial r^n}{\partial z}\right)$$

$$= \vec{i}\,nr^{n-1}\frac{\partial r}{\partial x} + \vec{j}\,nr^{n-1}\frac{\partial r}{\partial y} + \vec{k}\,nr^{n-1}\frac{\partial r}{\partial z}$$

$$= nr^{n-1}\left[\vec{i}\frac{x}{r} + \vec{j}\frac{y}{r} + \vec{k}\frac{z}{r}\right]$$

$$= \frac{nr^{n-1}}{r}\left[x\vec{i} + y\vec{j} + z\vec{k}\right]$$

$$= \frac{nr^{n-1}}{r}\vec{r}$$

5. Let us find a unit normal to the surface $x^2y + 2xz = 4$ at (2,-2,3).

Solution:

Given that $\phi = x^2y + 2xz$

$$\nabla\phi = \left(\vec{i}\frac{\partial}{\partial x} + \vec{j}\frac{\partial}{\partial y} + \vec{k}\frac{\partial}{\partial z}\right)(x^2y + 2xz)$$

$$= \vec{i}(2xy + 2z) + \vec{j}(x^2) + \vec{k}(2x)$$

At (2,-2,3):

$$\nabla\phi = \vec{i}(-8+6) + \vec{j}(4) + \vec{k}(4)$$

$$= -2\vec{i} + 4\vec{j} + 4\vec{k}$$

$$|\nabla\phi| = \sqrt{4+16+16} = \sqrt{36} = 6$$

Unit normal to the given surface at (2,-2,3):

$$\frac{\nabla\phi}{|\nabla\phi|} = \frac{2\vec{i} + 4\vec{j} + 4\vec{k}}{6}$$

$$= \frac{1}{3}\left(-\vec{i} + 2\vec{j} + 2\vec{k}\right)$$

6. Let us find the work done in moving a particle in the force field given by $\vec{F} = 3x^2\vec{i} + (2xz - y)\vec{j} + z\vec{k}$ along the straight line from (0, 0, 0) to (2, 1, 3).

Solution:

Let points be A (0, 0, 0) & B(2, 1, 3).

$$\bar{F}.d\bar{r} = 3x^2dx + (2xz - y)dy + zdz.$$

Equation of AB is:

$$\frac{x-0}{2-0} = \frac{y-0}{1-0} = \frac{z-0}{3-0}$$

$$\frac{x}{2}=\frac{y}{1}=\frac{z}{3}=t(\text{say})$$

$$x=2t\ y=t\ z=3t$$

$$dx=2dt\ dy=dt\ dz=3dt$$

At $(0, 0, 0)\ t=0$

At $(2, 1, 3)\ t=1$

$$\bar{F}.d\bar{r}=3(2t)^2(2dt)+[2(2t)(3t)-t](dt)+(3t)(3dt)$$

$$=6(4t^2)dt+(12t^2-t)dt+9t\,dt$$

$$=[24t^2+12t^2-t+9t]dt$$

$$=(36t^2+8t)dt.$$

Work done $=\int_C \bar{F}.d\bar{r}=\int_0^1(36t^2+8t)dt=[36t^3/3+8t^2/2]_0^1$

$$=12+4-0$$

$$=16$$

7. If $\vec{r}$ is the position vector of the point (x, y, z), let us prove that $\nabla^2 r^n = n(n+1)r^{n-2}$.

Solution:

Let, $\vec{r}=x\vec{i}+y\vec{j}+z\vec{k}$

Now, $\nabla r^n=\left(i\frac{\partial}{\partial x}+j\frac{\partial}{\partial y}+k\frac{\partial}{\partial z}\right)r^n$

$$=\sum i\frac{\partial}{\partial x}r^n=nr^{n-1}\sum i\frac{\partial r}{\partial x}$$

$$=nr^{n-1}\sum i\frac{x}{r}$$

$$=nr^{n-2}\,\vec{r}\qquad \left(\because \sum ix=\vec{r}\right)$$

Now, $\nabla^2 r^n=\nabla\cdot\nabla r^2$

$$= \nabla \cdot (nr^{n-2}\vec{r})$$

$$= n\nabla\left(r^{n-2}\right)\cdot \vec{r} + r^{n-2}\left(\nabla \cdot \vec{r}\right).$$

$$\nabla^2 r^n = n\left[(n-2)r^{n-4}\vec{r}\cdot\vec{r} + 3r^{n-2}\right]$$

$$= n\left[(n-2)r^{n-4}\, r^2 + 3r^{n-2}\right]$$

$$= n\left[(n-2)r^{n-2} + 3r^{n-2}\right]$$

$$= n(n+1)r^{n-2}$$

Unit Normal Vector to the Surface

If $\varphi(x, y, z)$ be a scalar function, then $\varphi(x, y, z) = c$ represents, a surface and the unit normal vector to the surface φ is given by:

$$\frac{\nabla\phi}{|\nabla\phi|}$$

Equation of the Tangent Plane and Normal to the Surface

Suppose $\vec{a}$ is the position vector of the point (x_0, y_0, z_0) on the surface $\phi(x,y,z) = c$. If $\vec{r} = x\vec{i} + y\vec{j} + z\vec{k}$ is the position vector of any point (x,y,z) on the tangent plane to the surface at $\vec{a}$, then the equation of the tangent plane to the surface φ at a given point $\vec{a}$ on it is given by $(\vec{r} - \vec{a}).\text{grad}\phi = 0$.

If $\vec{r}$ is the position vector of any point on the normal to the surface at the point $\vec{a}$ on it. The vector equation of the normal at a given point $\vec{a}$ on the surface f is $(\vec{r} - \vec{a}) \times \text{grad}\,\phi = 0$.

The Cartesian form of the normal at (x_0, y_0, z_0) on the surface:

$$\phi(x,y,z) = c$$

$$\frac{x - x_0}{\frac{\partial\phi}{\partial x}} = \frac{y - y_0}{\frac{\partial\phi}{\partial y}} = \frac{z - z_0}{\frac{\partial\phi}{\partial z}}$$

Directional Derivatives

The derivative of a point function (scalar or vector) in a particular direction is called its directional derivative along the direction. The directional derivative of a scalar point function ϕ in a given direction is the rate of change off in the direction. It is given by the ϕ component of grad ϕ in that direction.

The directional derivative of a scalar point function ϕ (x,y,z) in the direction of $\vec{a}$ is given by $\frac{\nabla\phi.\vec{a}}{|\vec{a}|}$.

Problem

1. Let us determine the directional derivative of $\phi = x^2yz + 4xz^2 + xyz$ at (1, 2, 3) in the direction of $2\vec{i} + \vec{j} - \vec{k}$.

Solution:

Given: $\phi = x^2yz + 4xz^2 + xyz$

$$\nabla\phi = \vec{i}\frac{\partial\phi}{\partial x} + \vec{j}\frac{\partial\phi}{\partial y} + \vec{k}\frac{\partial\phi}{\partial z}$$

$$= (2xyz + 4z^2 + xyz)\vec{i} + (x^2z + xz)\vec{j} + (x^2y + 8xz + xy)\vec{k}$$

$$= 54\vec{i} + 6\vec{j} + 28\vec{k}$$

$$\vec{a} = 2\vec{i} + \vec{j} - \vec{k}$$

$$|\vec{a}| = \sqrt{6}$$

$$D.D = \nabla\varphi.\frac{\vec{a}}{|\vec{a}|} = (54\vec{i} + 6\vec{j} + 28\vec{k}).\frac{2\vec{i} + \vec{j} - \vec{k}}{\sqrt{6}}$$

$$= \frac{1}{\sqrt{6}}(86)$$

Normal Derivative

Normal derivative= $|\nabla\varphi|$.

Problem

1. Let us calculate the greatest rate of increase of $\varphi = xyz^2$ at $(1,0,3)$.

Solution:

Given: $\varphi = xyz^2$

$$\varphi = xyz^2$$

$$\nabla\varphi = \vec{i}\frac{\partial\varphi}{\partial x} + \vec{j}\frac{\partial\varphi}{\partial y} + \vec{k}\frac{\partial\varphi}{\partial z}$$

$$\vec{i}(yz^2)+\vec{j}(xz^2)+\vec{k}(2xyz)=9\vec{j}$$

Greatest rate of increase $= |\nabla\varphi| = 9$

Divergence of a Vector

If $\vec{F}(x,y,z)$ is a continuously differentiable vector point function in a given region of space, then the divergences of $\vec{F}$ is defined by:

$$\nabla.\vec{F} = \operatorname{div}\vec{F} = \vec{i}\frac{\partial\vec{F}}{\partial x}+\vec{j}\frac{\partial\vec{F}}{\partial y}+\vec{k}\frac{\partial\vec{F}}{\partial z}$$

$$=\sum\vec{i}\frac{\partial\vec{F}}{\partial x}$$

If $\vec{F}=F_1\vec{i}+F_2\vec{j}+F_3\vec{k}$, then $\operatorname{div}\vec{F}=\nabla.\left(F_1\vec{i}+F_2\vec{j}+F_3\vec{k}\right)$

$$\text{i.e,} \operatorname{div}\vec{F}=\frac{\partial F_1}{\partial x}+\frac{\partial F_2}{\partial y}+\frac{\partial F_3}{\partial z}$$

Curl of Vector Function

If $\vec{F}$ (x, y, z) is a differentiable vector point function defined at each point (x, y, z), then the curl of $\vec{F}$ is defined by:

$$\operatorname{curl}\vec{F}=\nabla\times\vec{F}$$

$$=\vec{i}\times\frac{\partial\vec{F}}{\partial x}+\vec{j}\times\frac{\partial\vec{F}}{\partial y}+\vec{k}\times\frac{\partial\vec{F}}{\partial z}$$

$$=\sum\vec{i}\times\frac{\partial\vec{F}}{\partial x}$$

If $\vec{F}=F_1\vec{i}+F_2\vec{j}+F_3\vec{k}$, then curl $\vec{F}=\nabla\times\left(F_1\vec{i}+F_2\vec{j}+F_3\vec{k}\right)$

$$\operatorname{curl}\vec{F}=\begin{vmatrix} i & j & k \\ \frac{\partial}{\partial x} & \frac{\partial}{\partial y} & \frac{\partial}{\partial z} \\ F_1 & F_2 & F_3 \end{vmatrix}$$

$$=\vec{i}\left[\frac{\partial F_3}{\partial y}-\frac{\partial F_2}{\partial z}\right]-\vec{j}\left[\frac{\partial F_3}{\partial x}-\frac{\partial F_1}{\partial y}\right]+\vec{k}\left[\frac{\partial F_2}{\partial x}-\frac{\partial F_1}{\partial y}\right]$$ curl $\vec{F}$ is also said to be rotation $\vec{F}$.

Gradient:

- $\vec{\nabla}(f+g)=\vec{\nabla}f+\vec{\nabla}g$
- $\vec{\nabla}(cf)=c\vec{\nabla}f$, for any constant c
- $\vec{\nabla}(fg)=f\,\vec{\nabla}g+g\vec{\nabla}f$
- $\vec{\nabla}(f/g)=\left(g\vec{\nabla}f-f\vec{\nabla}g\right)/g^2$ at points $\vec{x}$ where $g(\vec{x})\neq o$
- $\vec{\nabla}(\vec{F}.\vec{G})=\vec{F}\times(\vec{\nabla}\times\vec{G})-(\vec{\nabla}\times\vec{F})\times\vec{G}+(\vec{G}.\vec{\nabla})\vec{F}+(\vec{F}.\vec{\nabla})\vec{G}$

Divergence:

- $\vec{\nabla}\cdot(\vec{F}+\vec{G})=\vec{\nabla}\cdot\vec{F}+\vec{\nabla}\cdot\vec{G}$
- $\vec{\nabla}\cdot(c\vec{F})=c\vec{\nabla}\cdot\vec{F}$, for any constant c
- $\vec{\nabla}\cdot(f\vec{F})=f\,\vec{\nabla}\cdot\vec{F}+\vec{F}\vec{\nabla}f$
- $\vec{\nabla}(\vec{F}\times\vec{G})=\vec{G}\cdot(\vec{\nabla}\times\vec{F})-\vec{F}\cdot(\vec{\nabla}\times\vec{G})$

Curl:

- $\vec{\nabla}\times(\vec{F}+\vec{G})=\vec{\nabla}.\vec{F}+\vec{\nabla}.\vec{G}$
- $\vec{\nabla}.(c\vec{F})=c\vec{\nabla}\times\vec{F}$, for any constant c
- $\vec{\nabla}.(f\vec{F})=f\,\vec{\nabla}\times\vec{F}+\vec{\nabla}f\times\vec{F}$
- $\vec{\nabla}(\vec{F}\times\vec{G})=\vec{F}.(\vec{\nabla}.\vec{G})-(\vec{\nabla}.\vec{F})\vec{G}+(\vec{G}.\vec{\nabla})\vec{F}-\vec{G}$

If ϕ is scalar function $\vec{A}$ is vector function $\vec{A}=a_1i+a_2j+a_3k$ then:

1. Grad $\phi=\nabla\phi=\frac{\partial\phi}{\partial x}i+\frac{\partial\phi}{\partial x}j+\frac{\partial\phi}{\partial x}k$

2. If $\vec{A}=a_1i+a_2j+a_3k$

$$\text{div}\,\vec{A}=\nabla.\vec{A}=\left(\frac{\partial}{\partial x}i+\frac{\partial}{\partial y}j+\frac{\partial}{\partial z}k\right).(a_1i+a_2j+a_3k)$$

$$\text{div}\,\vec{A}=\nabla.\vec{A}=\frac{\partial a_1}{\partial x}+\frac{\partial a_2}{\partial y}+\frac{\partial a_3}{\partial z}$$

3. If $\vec{A} = a_1 i + a_2 j + a_3 k$

$$\operatorname{div}\vec{A} = \nabla \times \vec{A} = \begin{pmatrix} i & j & k \\ \frac{\partial}{\partial x} & \frac{\partial}{\partial y} & \frac{\partial}{\partial z} \\ a_1 & a_2 & a_3 \end{pmatrix}$$

$$= \left(\frac{\partial a_3}{\partial y} - \frac{\partial a_2}{\partial z}\right) - j\left(\frac{\partial a_3}{\partial x} - \frac{\partial a_1}{\partial z}\right) + k\left(\frac{\partial a_2}{\partial x} - \frac{\partial a_1}{\partial y}\right)$$

4. Laplacian of $\phi = \frac{\partial^2 \phi}{\partial x^2} + \frac{\partial^2 \phi}{\partial y^2} + \frac{\partial^2 \phi}{\partial z^2}$

5. Laplacian of $\vec{A} = \nabla^2 \vec{A} = \frac{\partial^2 \vec{A}}{\partial x^2} + \frac{\partial^2 \vec{A}}{\partial y^2} + \frac{\partial^2 \vec{A}}{\partial z^2}$

5.1.2 Laplacian Operator

So far we have encountered $\nabla\phi = \left(\frac{\partial \phi}{\partial x}, \frac{\partial \phi}{\partial y}, \frac{\partial \phi}{\partial z}\right)$,

Which is a vector field formed from a scalar field and,

$$\nabla . A = \left(\frac{\partial A_x}{\partial x} + \frac{\partial A_y}{\partial y} + \frac{\partial A_z}{\partial z}\right)$$

Which is a scalar field formed from a vector field. There are two ways in which we can combine gradient and divergence. We can either form the vector field $\nabla(\nabla . A)$ or the scalar field $\nabla.(\nabla\phi)$. The former is not particularly interesting, but the scalar field $\nabla.(\nabla\phi)$ turns up in a great many physical problems.

Let us introduce the heat flow vector h, which is the rate of flow of heat energy per unit area across a surface perpendicular to the direction of h. In many substances, the heat flows directly down the temperature gradient, so that we can write:

$$h = -k\nabla T,$$

Where k is the thermal conductivity. The net rate of heat flow $\oint_s h.dS$ out of some closed surface S must be equal to the rate of decrease of heat energy in the volume V enclosed by S. Thus, we have:

$$\oint_s h.dS = -\frac{\partial}{\partial t}\left(\int cT\,dv\right)$$

Where c is the specific heat. It follows from the divergence theorem that:

$$\nabla.h = -c\frac{\partial T}{\partial t}$$

Taking the divergence of both sides of equation, we obtain:

$$\nabla.(k\nabla T) = c\frac{\partial T}{\partial t}$$

If k is constant then the above equation can be written:

$$\nabla.(\nabla T) = \frac{c}{k}\frac{\partial T}{\partial t}$$

The scalar field $\nabla.(\nabla T)$ takes the form:

$$\nabla.(\nabla T) = \frac{\partial}{\partial x}\left(\frac{\partial T}{\partial x}\right) + \frac{\partial}{\partial y}\left(\frac{\partial T}{\partial y}\right) + \frac{\partial}{\partial z}\left(\frac{\partial T}{\partial z}\right)$$

$$= \frac{\partial^2 T}{\partial x^2} + \frac{\partial^2 T}{\partial y^2} + \frac{\partial^2 T}{\partial Z^2} = \nabla^2 T$$

Here, the scalar differential operator is called the Laplacian. The Laplacian is a good scalar operator because it is formed from a combination of divergence and gradient.

$$\nabla^2 \equiv \frac{\partial^2}{\partial x^2} + \frac{\partial^2}{\partial y^2} + \frac{\partial^2}{\partial z^2}$$

In one dimension, $\nabla^2 T$ reduces to $\nabla^2 T / \partial x^2$. Now, $\partial^2 T / \partial x^2$ is positive if T(x) is concave and negative if it is convex. So, if T is less than the average of T in its surroundings then $\nabla^2 T / \partial y^2$ is positive and vice versa.

In two dimensions:

$$\nabla^2 T = \frac{\partial^2 T}{\partial x^2} + \frac{\partial^2 T}{\partial y^2}$$

Consider a local minimum of the temperature. At the minimum, the slope of T increases in all directions, so $\nabla^2 T$ is positive. Likewise, $\nabla^2 T$ is negative at a local maximum. Now, consider the steep-sided valley in T. Suppose that the bottom of the valley runs parallel to the x-axis. At the bottom of the valley $\nabla^2 T / \partial y^2$ is large and positive, whereas

$\nabla^2 T / \partial x^2$ is small and may even be negative. Thus, $\nabla^2 T$ is positive and this is associated with T being less than the average local value.

5.1.3 Line Integral and Work Done

Problems

1. If $\vec{F} = (3x^2 + 6y)\vec{i} - 14yz\,\vec{j} + 20xz^2\vec{k}$, let us Evaluate $\int_c \vec{F}.dr$ from (0, 0, 0) to (1,1,1) along the curve x = t, y = t^2, z = t^3.

Solution:

The end points are (0, 0, 0) and (1, 1, 1).

These points correspond to t = 0 and t = 1.

$$\therefore dx = dt, dy = 2t, dz = 3t^2$$

$$\int_c \vec{F}.dr = (3x^2 + 6y)dx - 14yzdy + 20xz^2dz$$

$$= \int_0^1 (3t^2 + 6t^2)dt - 14t^5(2tdt) + 20t^7(3t^2)dt$$

$$= \int_0^1 (9t^2 - 28t^6 + 60t^9)dt$$

$$= (3t^3 - 4t^7 + 6t^{10})_0^1$$

$$= [(3 - 4 + 6) - 0] = 5$$

2. Let us show that $\vec{F} = x^2\vec{i} + y^2\vec{j} + z^2\vec{k}$ is a conservative vector field.

Solution:

If $\vec{F}$ is conservation then $\nabla \times \vec{F} = 0$,

$$\nabla \times \vec{F} = \begin{vmatrix} \vec{i} & \vec{j} & \vec{k} \\ \frac{\partial}{\partial x} & \frac{\partial}{\partial y} & \frac{\partial}{\partial z} \\ x^2 & y^2 & z^2 \end{vmatrix} = 0\vec{i} + 0\vec{j} + 0\vec{k} = 0$$

$\therefore \vec{F}$ is a conservative vector field.

Work Done

Let F be a vector field and C be a curve defined by a vector valued function r. Then the work done by F on an object moving along C is given by:

$$\text{Work} = \int_c F.dr = \int_a^b F(x(t)), y(t), z(t).r'(t)dt$$

3. Find the work done by the vector field F(x,y,z) = xi + 3xyj - (x + z)k on a particle moving along the line segment that goes from (1,4,2) to (0,5,1).

Solution:

We first have to parameterize the curve. We have:

$$r(t) = <1,4,2> + [<0,5,1> - <1,4,2>]t = <1-t,\ 4+t,\ 2-t>$$

and

$$r'(t) = -i + j - k$$

Taking the dot product, we get:

$$F.r'(t) = -x + 3xy + x + z = 3xy + z$$

$$= 3(1-t)(4+t) + (2-t) = -3t^2 - 10t + 14$$

Now we just integrate:

$$\int_0^1 (-3t^2 - 10t + 14)dt = \left[-t^3 - 5t^2 + 14t\right]_0^1 = 8$$

4. If $\vec{F} = (3x^2 + 6y)\vec{i} - 14yz\vec{j} + 20xz^2\vec{k}$, let us Evaluate $\int_c \vec{F}.dr$ from (0,0,0) to (1,1,1) along the curve x = t, y = t², z = t³.

Solution:

The end points are (0, 0, 0) and (1, 1, 1).

These points correspond to t = 0 and t = 1.

$$\therefore dx = dt, dy = 2t, dz = 3t^2$$

$$\int_c \vec{F}.dr = (3x^2 + 6y)dx - 14yzdy + 20xz^2dz$$

$$= \int_0^1 (3t^2 + 6t^2)dt - 14t^5(2tdt) + 20t^7(3t^2)dt$$

$$=\int_0^1 \left(9t^2 - 28t^6 + 60t^9\right)dt$$

$$=\left(3t^3 - 4t^7 + 6t^{10}\right)_0^1$$

$$=\left[(3-4+6)-0\right]=5$$

5. Let us show that $\vec{F} = x^2\vec{i} + y^2\vec{j} + z^2\vec{k}$ is a conservative vector field.

Solution:

If $\vec{F}$ is conservation then $\nabla \times \vec{F} = 0$

$$\nabla \times \vec{F} = \begin{vmatrix} \vec{i} & \vec{j} & \vec{k} \\ \frac{\partial}{\partial x} & \frac{\partial}{\partial y} & \frac{\partial}{\partial z} \\ x^2 & y^2 & z^2 \end{vmatrix} = 0\vec{i} + 0\vec{j} + 0\vec{k} = 0$$

$\therefore \vec{F}$ is a conservative vector field.

Notice that work done by a force field on an object moving along a curve depends on the direction that the object goes. In fact the opposite direction will produce the negative of the work done in the original direction. This is clear from the fact that everything is the same except the order which we write a and b.

5.1.4 Surface Integrals and Volume Integral

In ordinary calculus we compute integrals of real functions of a real variable, that is, we compute integrals of functions of the type.

$$y = f(x)$$

Where x and y are real numbers.

In vector analysis we compute integrals of vector functions of a real variable; that is we compute integrals of functions of the type.

$$f(t) = f1(t)i + f2(t)j + f3(t)k$$

Or equivalently:

$$\overline{F} = \begin{bmatrix} f_1(t) \\ f_2(t) \\ f_3(t) \end{bmatrix}$$

Where $f_1(t)$, $f_2(t)$, and $f_3(t)$ are real functions of the real variable t.

Surface Integrals

Definition: Consider a surface S. Let n denote the unit outward normal to the surface S. Let R be the projection of the surface x on the XY plane. Let $\vec{f}$ be a vector valued defined in some region containing the surface S. Then the surface integral of $\vec{f}$ is defined to be

$$\iint_S \vec{f}.\hat{n}ds = \iint_R \frac{\vec{f}.\hat{n}}{|\vec{n}.\vec{k}|}dx.dy .$$

Problems

1. Let us evaluate $\iint_S \vec{F}.\hat{n}\,ds$ where $\vec{F} = 18z\vec{i} - 12j + 3y\vec{k}$ as S is the part of the plane 2x+3y+6z=12 which is in the first octant.

Solution:

Given:

$$\vec{F} = 18z\vec{i} - 12j + 3y\vec{k}$$

Let, $\phi = 2x + 3y + 6z - 12$

$$\hat{n} = \frac{\nabla\phi}{|\nabla\phi|}$$

$$\nabla\phi = \vec{i}\frac{\partial\phi}{\partial x} + \vec{j}\frac{\partial\phi}{\partial y} + \vec{k}\frac{\partial\phi}{\partial z}$$

$$= 2\vec{i} + 3\vec{j} + 6\vec{k}$$

$$|\nabla\phi| = \sqrt{4+9+36} = \sqrt{49} = 7$$

$$\hat{n} = \frac{\nabla\phi}{|\nabla\phi|} = \frac{2\vec{i} + 3\vec{j} + 6\vec{k}}{7}$$

$$\vec{F}\cdot\hat{n} = (18z\vec{i} - 12\vec{j} + 3y\vec{k}).\left(\frac{2\vec{i} + 3\vec{j} + 6\vec{k}}{7}\right)$$

$$= \frac{36z - 36 + 18y}{7}$$

$$= \frac{1}{7}\left[36\left(\frac{12-2x-3y}{6}\right) - 36 + 18y\right] \quad \left[\because z = \frac{12-2x-3y}{6}\right]$$

$$= \frac{1}{7}\left[6(12-2x-3y)-36+18y\right] = \frac{1}{7}[36-12x]$$

$$\iint_S \vec{F}\cdot\hat{n}\,ds = \iint_R \frac{36-12x}{7}\frac{dx\,dy}{|\hat{n}.\hat{k}|}$$

$$\hat{n}\cdot\vec{k} = \frac{2\vec{i}+3\vec{j}+6\vec{k}}{7}.\vec{k} = \frac{6}{7}$$

$$\therefore \iint_S \vec{F}.\hat{n}\,ds = \iint_R \frac{36-12x}{7}\frac{7}{6}\,dx\,dy$$

$$= \iint_R (6-2x)\,dx\,dy$$

Given plane equation is $2x+3y+6z=12$ in XOY area z=0

$$2x+3y=12$$

$$3y = 12-2x \qquad \left(\therefore y \text{ varies } 0 \text{ to } \frac{12-2x}{3} \text{ in X plane } y=0, z=0\right)$$

$$y = \frac{12-2x}{3} \qquad \therefore 2x = 12$$

$$x = 6$$

$\therefore$ x varies from 0 to 6

$$= \int_0^6 \int_0^{\frac{12-2x}{3}} (6-2x)\,dy\,dx$$

$$= \int_0^6 (6y-2xy)_0^{\frac{12-2x}{3}}\,dx$$

$$= \int_0^6 \left[6\left(\frac{12-2x}{3}\right) - 2x\left(\frac{12-2x}{3}\right)\right]dx$$

$$= \int_0^6 \left(24-4x-8x+\frac{4x^2}{3}\right)dx$$

$$= \int_0^6 \left(24-12x+\frac{4x^2}{3}\right)dx$$

$$= \left[24-6x^2+\frac{4x^3}{9}\right]_0^6 = 24$$

2. Let us evaluate $\iint_S \vec{f}.\hat{n}ds$ where $\vec{F} = z\vec{i} + x\vec{j} - y^2 z\vec{k}$ and S is the surface of the cylinder $x^2+y^2=1$ including in the first octant between the planes z=0 and z=2.

Solution:

Given; $\vec{F} = z\vec{i} + x\vec{j} - y^2 z\vec{k}$

$$\phi = x^2 + y^2 - 1$$

$$\nabla\phi = 2x\vec{i} + 2y\vec{j}$$

$$|\nabla\phi| = \sqrt{4x^2 + 4y^2}$$

$$= 2\sqrt{x^2 + y^2} = 2$$

The unit normal $\hat{n}$ to the surface $= \dfrac{\nabla\phi}{|\nabla\phi|}$

$$= \frac{2xi + 2yj}{2} = xi + yj$$

$$\vec{F}.\hat{n} = \left(z\vec{i} + x\vec{j} - y^2 z\vec{k}\right).\left(x\vec{i} + y\vec{j}\right) = xz + xy$$

Now, $\iint_S \vec{F}.\hat{n}\,ds = \hat{n}\,ds = \iint_R \vec{F}\cdot\hat{n}\dfrac{dy\,dz}{|\hat{n}\cdot\vec{t}|}$

Where R is the projection of S on YZ plane.

$$= \iint_R (xz + xy)\frac{dy\,dz}{x} \qquad \left[\because \hat{n}.\hat{i} = \left(x\hat{i} + y\hat{j}\right)\hat{i} = x\right]$$

$$= \iint_R (2 + y)\,dy\,dz$$

$$= \int_0^2\int_0^1 (2 + y)\,dy\,dz$$

$$= \int_0^2 \left[zy + \frac{y^2}{2}\right]_0^1 dz = \int_0^2 \left(x + \frac{1}{2}\right)dz$$

$$= \left[\frac{z^2}{2} + \frac{1}{2}z\right]^2$$

$$=\left(\frac{4}{2}+\frac{3}{2}\right)-(0+0)$$

$$=(2+1)=3$$

Volume Integral

The volume integral of F(x,y,z) over region enclosing a volume V is given by $\iiint_V F(x,y,z)dV$ or $\iiint_V F(x,y,z)dx\,dy\,dz$.

Problem

1. If $\vec{F}=(2x^2-3x)\vec{i}-2xy\,\vec{j}-4x\,\vec{k}$., let us evaluate $\iiint_V \nabla\times\vec{F}dv$ where V is the region bounded by x=0,y=0,z=0 and 2x+2y+z=4.

Solution:

$$\nabla\times\vec{F}=\begin{vmatrix}\vec{i} & \vec{j} & \vec{k}\\ \frac{\partial}{\partial x} & \frac{\partial}{\partial y} & \frac{\partial}{\partial z}\\ 2x^2-3z & -2xy & -4x\end{vmatrix}$$

$$=\vec{i}(0)+\vec{j}(-3+4)+\vec{k}(-2y-0)$$

$$=\vec{j}-2y\vec{k}$$

$$\therefore\iiint_V \nabla\times\vec{F}dV=\int_0^2\int_0^{2-x}\int_0^{4-2x-2y}(\vec{j}-2y\vec{k})dz\,dy\,dx$$

$$=\int_0^2\int_0^{2-x}(z\vec{j}-2yz\vec{k})_0^{4-2x-2y}dy\,dx$$

$$=\int_0^2\int_0^{2-x}[4-2x-2y]\vec{j}-2y(4-2x-2y\vec{k})dy\,dx$$

$$=\int_0^2\left[(4y-2xy-y^2)\vec{i}-\left(4y^2-2xy^2-\frac{4y^3}{3}\right)k\right]_0^{2-x}dx$$

$$=\int_0^2\left[\left[4(2-x)-2x(2-x)-(2-x)^2\right]j-\left[4(2-x)^2-2x(2-x)^2-\frac{4}{3}(2-x)^3\right]k\right]dx$$

$$=\int_0^2(8-4x-4x+2x^2-4-x^2+4x)\vec{j}-\left[(16-16x+4x^2-8x+8x_n^2-2x^3)\right.$$

$$\left.-\frac{4}{3}(8-12x+6x^2-x^3)k\right]dx$$

$$=\int_0^2\left[(4-4x+x^2)\right]\vec{j}-\frac{\vec{k}}{3}\left[16-24x+12x^2-2x^3\right]dx$$

$$=\left[\left(4x-2x^2+\frac{x^3}{3}\right)\right]\vec{j}-\frac{\vec{k}}{3}\left[16x-12x^2+4x^3-\frac{x^4}{2}\right]_0^2$$

$$=\left(8-8+\frac{8}{3}\right)\vec{j}-\frac{\vec{k}}{3}(32-48+32-8)=\frac{8}{3}(\vec{j}-\vec{k})$$

5.2 Green's Theorem

Statement: If M(x,y) and N(x,y) are continuous functions with continuous partial derivatives in a region R of the xy plane bounded by a simple closed curve C, then $\oint_c Mdx+ndy=\iint_R\left(\frac{\partial N}{\partial x}-\frac{\partial M}{\partial y}\right)dxdy$ where C is the curve described in the positive direction.

Problems

1. Let us verify Green's theorem in the plane for $\int_C(3x^2-8y^2)dx+(4y-6xy)dy$ where C is the boundary of the region define by x = 0, y = 0, x + y = 1.

Solution:

Green's Theorem gives:

$$\int_C Mdx+Ndy=\iint_R\left(\frac{\partial N}{\partial x}-\frac{\partial M}{\partial y}\right)dxdy$$

Where R is the region bounded by C and C is taken in anti-clockwise direction.

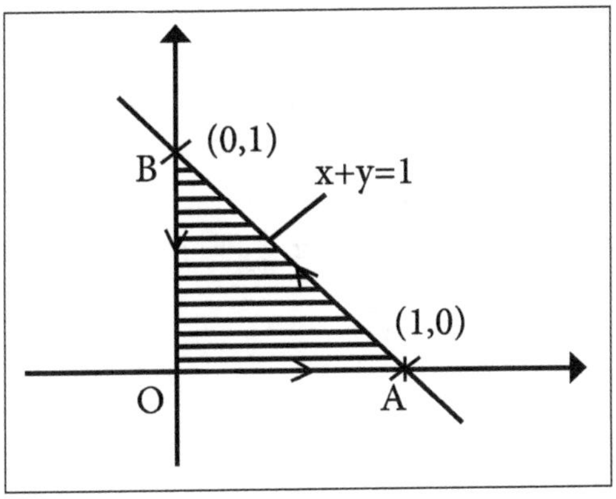

From the given integral:

$$M = 3x^2 - 8y^2 N = 4y - 6xy$$

LHS:

Let, $I_1 = \int_C Mdx + Ndy$

$$I_1 = \int_C = \int_{OA} + \int_{AB} + \int_{BO}$$

Along OA, y = 0 and x : 0 to 1

$$dy = 0$$

$$\therefore Mdx + Ndy = 3x^2 \ dx$$

$$\therefore OA = \int_0^1 3x^2 dx = 3\left[\frac{x^3}{3}\right]_0^1 = (1-0) = 1$$

Along AB, $x + y = 1$. x varies from 1 to 0 and y varies from 0 to 1.

$$Mdx + Ndy = \left[\left(3x^2 - 8(1-x)^2\right)\right]dx + \left(4y - 6y(1-y)\right)dy$$

$$= \left[\left(3x^2 - 8(1 + x^2 - x)\right]dx + \left[4y - 6y + 6y^2\right]dy$$

$$= \left[3x^2 - 8 - 8x^2 + 16x\right]dx + \left[6y^2 - 2y\right]dy$$

$$= \left[16x - 5x^2 - 8\right]dx + \left[6y^2 - 2y\right]dy$$

$$\therefore \int_{AB} = \int_{AB} \left(16x - 5x^2 - 8\right)dx + \left(6y^2 - 2y\right)dy$$

$$= \int_1^0 \left(16x - 5x^2 - 8\right)dx + \int_0^1 \left(6y^2 - 2y\right)$$

$$= \left[16\frac{x^2}{2} - 5\frac{x^3}{3} - 8x\right]_1^0 + \left[2\frac{y^3}{1} - 2\frac{y^2}{2}\right]_0^1$$

$$= 0 - \left(8 - \frac{5}{3} - 8\right) + \left[(2-1) - 0\right]$$

$$= \frac{5}{3} + 1 = \frac{8}{3}$$

Along BO:

$x = 0$ and $y : 1$ to 0

$Mdx + Ndy = 6 + (4y)dy = 4y\,dy$

$$\therefore \int_{BO} = \int_0^0 4y\,dy = 4\left[\frac{y^2}{2}\right]_1^0 = 2(0-1)$$

$$\therefore I_1 = 1 + \frac{8}{3} - 2 = \frac{3+8-6}{3} = \frac{5}{3}$$

$$I_1 = \frac{5}{3}$$

RHS:

$$\text{Let} \quad I_2 = \iint_R \left(\frac{\partial N}{\partial x} - \frac{\partial M}{\partial y}\right)dxdy$$

$$\frac{\partial N}{\partial x} = -6y, \quad \frac{\partial M}{\partial y} = -16y$$

$$\therefore \frac{\partial N}{\partial x} - \frac{\partial M}{\partial y} = -6y - (-16y) = 10y$$

$$\therefore I_2 = \int_0^1 \int_0^{1-x} 10y\,dy\,dx = 10\int_0^1 \left[\frac{y^2}{2}\right]_0^{1-x} dx$$

$$= 5\int_0^1 (1-x)^2\,dx = 5\left[\frac{(1-x)^3}{-3}\right]_0^1 = \frac{-5}{3}[0-1]$$

$$I_2 = \frac{5}{3}$$

$I_1 = I_2.$

$\Rightarrow$ LHS = RHS

$\therefore$ Green's theorem is verified.

2. Let us verify Green's theorem for $\int_C (3x^2 - 8y^2)dx + (4y - 6xy)dy$ where C is the boundary of the region defined by $x = y^2$, $y = x^2$.

Solution:

Let $I = \int_C (3x^2 - 8y^2)dx + (4y - 6xy)dy$

Green's theorem states that:

$$\int_C (Pdx + Qdy) = \iint_R \left(\frac{\partial Q}{\partial x} - \frac{\partial P}{\partial y}\right) dx\, dy \quad ...(1)$$

Here $P = 3x^2 - 8y^2; Q = 4y - 6xy$

$$\frac{\partial P}{\partial y} = -16y; \ \frac{\partial P}{\partial y} = -6y$$

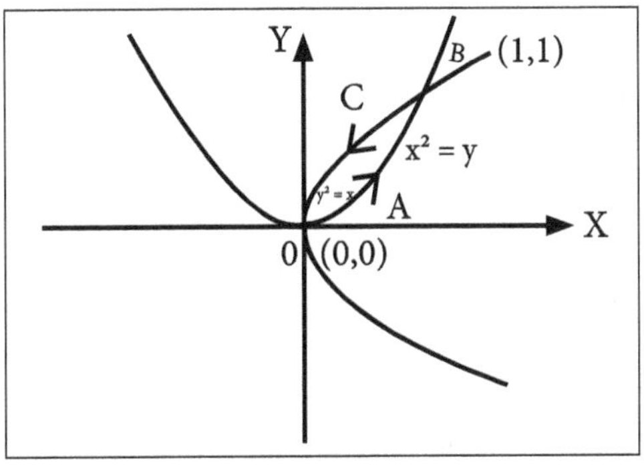

LHS:

$$\int_C (Pdx + Qdy) = \int_{OAB} = \int_{BCO} = I_1 + I_2$$

I_1 along the curve OAB ; $x^2 = y \Rightarrow 2xdx = dy$

$$I_1 = \int_0^1 \left[(3x^2 - 8x^4)dx + (4x^2 - 6x^3)2xdx\right]$$

$$= \int_0^1 (3x^2 + 8x^3 - 20x^4)dx = (x^3 + 2x^4 - 4x^5)_0^1$$

$$I_1 = -1$$

I_2 along the curve BCO ; $x = y^2$; $dx = 2ydy$

$$I_2 = \int_1^0 \left[\left(3y^4 - 8y^2\right)2ydy + \left(4y - 6y^3\right)dy\right]$$

$$= \int_1^0 \left[6y^5 - 22y^3 + 4y\right] dy$$

$$= \left(y^6 - \frac{11y^4}{2} + 2y^2\right)_1^0$$

$$I_2 = \frac{5}{2}$$

$$\text{L.H.S} = I_1 + I_2 = -1 + \frac{5}{2} = \frac{3}{2} \qquad \text{...(I)}$$

$$\text{R.H.S} = \iint_R \left(\frac{\partial Q}{\partial x} - \frac{\partial P}{\partial y}\right) dx\,dy$$

$$= \iint_R 10y\,dx\,dy = 10\iint y\,dy\,dx$$

Limit for dy dx: y varies from x^2 to $\sqrt{x}$

x varies from 0 to 1:

$$= 10\int_0^1 \int_{x^2}^{\sqrt{x}} y\,dy\,dx$$

$$= 10\int_0^1 \left(\frac{y^2}{2}\right)_{x^2}^{\sqrt{x}} dx = 5\int_0^1 \left[x - x^4\right] dx$$

$$= 5\left[\frac{x^2}{2} - \frac{x^5}{5}\right]_0^1 = \frac{3}{2} \qquad \text{...(II)}$$

From (I) & (II):

$$\int_C Pdx + Qdy = \iint_R \left(\frac{\partial Q}{\partial x} - \frac{\partial P}{\partial y}\right) dx\,dy = \frac{3}{2}$$

3. Let us verify Green's theorem in a plane for: $\int_C \left[\left(3x^2 - 8y^2\right)dx + \left(4y - 6xy\right)dy\right]$ where C is the boundary of the region defined by x = 0, y = 0 and x + y = 1.

Solution:

Green's Theorem gives:

$$\int_C M dx + N dy = \iint_R \left(\frac{\partial N}{\partial x} - \frac{\partial M}{\partial y} \right) dx\, dy$$

C is the boundary of the region R defined by x = 0, y = 0 & x + y = 1.

LHS:

Let:

$$I_1 = \int_C M dx + N dy$$

$$= \int_C$$

$$= \int_{OA} + \int_{AB} + \int_{BO}$$

$$M dx + N dy = \left(3x^2 - 8y^2\right) dx + \left(4y - 6xy\right) dy$$

$$\therefore \; M = 3x^2 - 8y^2, \; N = 4y - 6xy$$

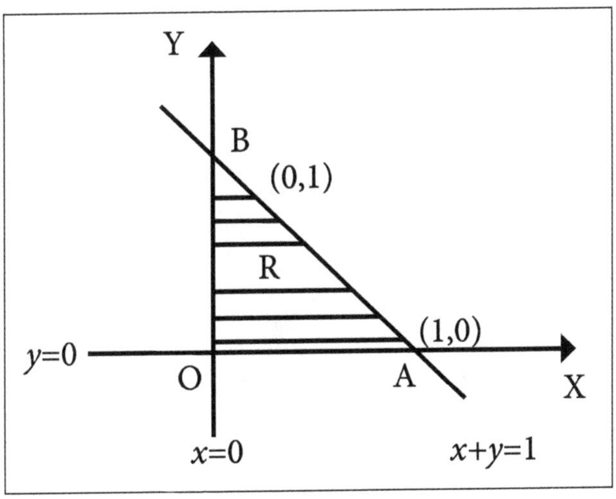

Along OA:

$$y = 0$$

$$\Rightarrow dy = 0 \text{ and x varies from 0 to 1}$$

$$\therefore \; M dx + N dy = \left(3x^2 - 0\right) dx + 0$$

$$= 3x^2 dx$$

$$\int_{OA} = \int_0^1 3x^2\,dx = 3\left[\frac{x^3}{3}\right]_0^1 = 1-0=1$$

Along AB:

$$x+y=1\quad y=1-x$$
$$x=1-y$$

x varies from 1 to 0.

y varies from 0 to 1.

$$\begin{aligned} Mdx+Ndy &= \left[3x^2-8(1-x)^2\right]dx+\left[4y-6(1-y)y\right]dy \\ &= \left[3x^2-8(1+x^2-2x)\right]dx+\left[4y-6(y-y^2)\right]dy \\ &= \left[3x^2-8-8x^2+16x\right]dx+\left[4y-6y+6y^2\right]dy \\ &= \left[-5x^2+16x-8\right]dx+\left[6y^2-2y\right]dy. \end{aligned}$$

$$\therefore \int_{AB} = \int_{AB}(-5x^2+16x-8)dx+(6y^2-2y)dy$$

$$= \int_1^0(-5x^2+16x-8)dx+\int_1^0(6y^2-2y)dy$$

$$= \left[-\frac{5x^3}{3}+\frac{16x^2}{2}-8x\right]_1^0+\left[\frac{6y^3}{3}-\frac{7y^2}{2}\right]_0^1$$

$$= 0+0-0-\left\{-\frac{5}{3}+8-8\right\}+2-1-0-0$$

$$= \frac{5}{3}+1=\frac{8}{3}$$

Along BO:

$$x=0\quad dx=0$$

y varies from 1 to 0.

$$Mdx+Ndy-0+(4y)dy$$

$$\therefore \int_{BO} = \int_1^0 4y\,dy = 4\left[\frac{y^2}{2}\right]_1^0 = 2(0-1) = -2$$

$$= -2$$

$$\therefore LHS = I_1 = 1 + \frac{8}{3} - 2$$

$$= \frac{5}{3}$$

RHS:

Let,

$$I_2 = \iint_R \left(\frac{\partial N}{\partial x} - \frac{\partial M}{\partial y} \right) dx\, dy$$

$$M = 3x^2 - 8y^2,\ N = 4y - 6xy$$

$$\frac{\partial M}{\partial y} = 0 - 16y,\ \frac{\partial N}{\partial x} = 0 - 6y$$

$$\therefore \left(\frac{\partial N}{\partial x} - \frac{\partial M}{\partial y} \right) = -6y - (-16y) = 10y$$

Limits are:

$x : 0$ to $1 - y$.

(Constant) $y : 0$ to 1.

$$\therefore I_2 = \int_0^1 \int_0^{1-y} 10y\, dx\, dy = \int_0^1 [x]_0^{1-y}\ 10y\, dy$$

$$= 10 \int_0^1 (1-y) y\ dy = 10 \int_0^1 (y - y^2) dy$$

$$= 10 \left[y^2/2 - y^3/3 \right]_0^1 = 10 \left[(1/2 - 1/3) - 0 \right] = 10(1/6)$$

$$= 5/3$$

∴ LHS = RHS. Hence Green's theorem is verified.

4. Let us verify Green's theorem in plane for $\int_C \left[(3x^2 - 8y^2) dx + (4y - 6x) dy \right]$ where C is the boundary of the region bounded by the lines x = 0, y = 0 and x + y = 1.

Solution:

Greens theorem states that:

$$\int_C (Pdx + Qdy) = \iint_R \left(\frac{\partial Q}{\partial x} - \frac{\partial P}{\partial y} \right) dx\, dy$$

Given : $\int_C \left[\left(3x^2-8y^2\right)dx+\left(4y-6xy\right)dy\right]$

Here, $P=\left(3x^2-8y^2\right)$; $Q=\left(4y-6xy\right)$

$$\frac{\partial P}{\partial y}=-16y\ ;\ \frac{\partial Q}{\partial x}=-6y$$

The region:

L.H.S.,

$$\int_C \left[\left(3x^2-8y^2\right)dx+\left(4y-6xy\right)dy\right]$$

$$=\int_{OA}+\int_{AB}+\int_{BO}$$

$$=I_1+I_2+I_3$$

I_1 along the line $y=0$; $dy=0$

$$I_1=\int_0^1 3x^2\,dx=\left(x^3\right)_0^1=1$$

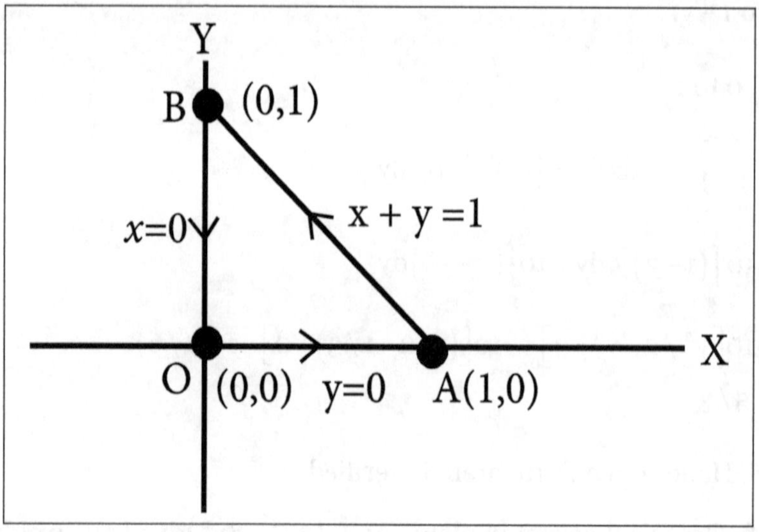

Here, $x=1-y$ or $y=1-x$ and $dx=-dy$

$$I_2=\int_1^0\left[3x^2-8\left(1-2x+x^2\right)\right]dx+\int_0^1\left[4y-6y\left(1-y\right)\right]dy$$

$$=\int_1^0\left[-5x^2+16x-8\right]dx+\int_0^1\left[6y^2-2y\right]dy$$

$$=\frac{8}{3}$$

I_3 = along the line $x=0$; $dx=0$

$$I_3=\int_1^0 4y\,dy=\left(2y^2\right)_0^1=0$$

$$LHS=I_1+I_2+I_3=1+\frac{8}{3}-2=\frac{5}{3} \qquad ...(I)$$

$$R.H.S=\iint_R 10\,y\,dxdy=\int_0^1\int_0^{1-y}10y\,dx\,dy$$

$$=10\int_0^1 y\left(x\right)_0^{(1-y)}dy=10\int_0^1\left[y-y^2\right]dy$$

$$=10\left[\frac{y^2}{2}-\frac{y^3}{3}\right]_0^1=10\left[\frac{1}{2}-\frac{1}{3}\right]$$

$$=\frac{5}{3} \qquad ...(II)$$

From (I) and (II);

$$\int_C Pdx+Qdy=\iint_R\left(\frac{\partial Q}{\partial x}-\frac{\partial P}{\partial y}\right)dxdy=\frac{5}{3}$$

Hence Green's theorem is verified.

5. Let us verify Green's theorem for $\bar{v}=\left(x^2+y^2\right)\bar{i}-2xy\bar{j}$ taken around the rectangle bounded by the lines x = ± a, y = 0 and y = b.

Solution:

Green's theorem states that:

$$\int_C\left(Pdx+Qdy\right)=\iint_R\left[\frac{\partial Q}{\partial x}-\frac{\partial P}{\partial y}\right]dx\,dy \qquad ...(1)$$

Here, $\int\overline{V}.\overline{dr}=\int\left(x^2+y^2\right)dx-2xy\,dy$

Here, $p=x^2+y^2$; $Q=-2xy$

$$\frac{\partial P}{\partial y}=2y \;;\; \frac{\partial Q}{\partial x}=-2y$$

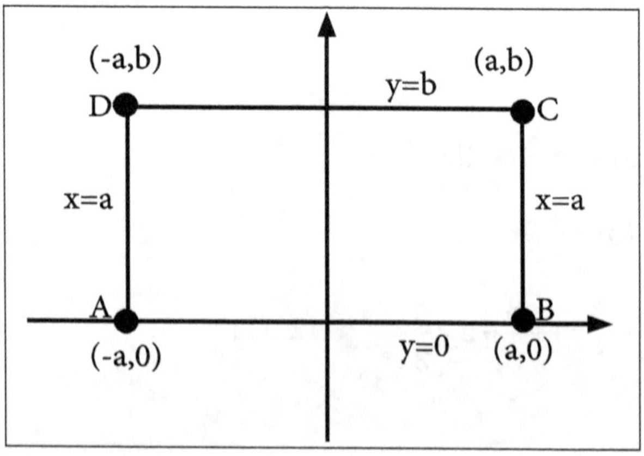

LHS of Green's Theorem:

$$\int (x^2+y^2)dx - 2xy\,dy = \int_{AB} + \int_{BC} + \int_{CD} + \int_{DA}$$

$$= I_1 + I_2 + I_3 + I_4$$

I_1: along the line y = 0; dy = 0.

$$I_1 = \int_{AB} x^2 dx = \int_{-a}^{a} x^2\,dx = 2\int_{0}^{a} x^2\,dx = \left(\frac{2x^3}{3}\right)_0^a$$

$$= \frac{2a^3}{3}$$

I_2: along the line x = a; dx = 0.

$$I_2 = \int_{BC} -2ay\,dy = -2a\int_{0}^{b} y\,dy$$

$$= -2a\left(\frac{y^2}{2}\right)_0^b = -ab^2$$

I_3: along the line y = b; dy = 0.

$$I_2 = \int_{CD} (x^2+b^2)dx = \int_{a}^{-a} (x^2+b^2)dx$$

$$= \left[\frac{x^3}{3} + b^2x\right]_a^{-a} = \left[\left(-\frac{a^3}{3} - ab^2\right) - \left(\frac{a^3}{3} + ab^2\right)\right]$$

$$= -\frac{2a^3}{3} - 2ab^2$$

I_4: along the line x = – a; dx = o

$$I_2 = \int_{DA} 2ay\, dy = 2a \int_b^o y\, dy$$

$$= a\left(y^2\right)_b^o = -ab^2$$

∴ L.H.S of Green's theorem.

$$\int_C \left(x^2 + y^2\right)dx - 2xy\,dy = \frac{2a^3}{3} - ab^2 - \frac{2a^3}{3} - 2ab^2 - ab^2$$

$$= -4ab^2 \qquad ...(I)$$

R.H.S of Green's theorem:

$$\iint_R \left[\frac{\partial Q}{\partial x} - \frac{\partial P}{\partial y}\right] dx\,dy = \iint_R [-2y - 2y]\,dx\,dy$$

Limit for dx dy:

x varies from - a to a:

y varies from o to b:

$$= \int_o^b \int_{-a}^a -4y\,dx\,dy$$

$$= \int_o^b -4y(x)_{-a}^a\, dy$$

$$= -4\int_o^b y[a + a]\,dy$$

$$= -8a\int_o^b y\,dy = -8a\left(\frac{y^2}{2}\right)_o^b$$

$$= -4ab^2 \qquad ...(II)$$

From (I) and (II), Green's theorem is verified.

5.2.1 Stoke's Theorem and Gauss's Divergence Theorems

Gauss's Divergence Theorem

Statement: The surface integral of the normal component of a vector function F over a closed surface S enclosing volume V is equal to the volume integral of the divergence of

F taken throughout the volume V:

$$\iint_S \vec{F}\cdot\hat{n}\,ds = \iiint_V \nabla\cdot\vec{F}\,dv$$

Problems

1. Let us verify Gauss-divergence theorem for the vector function $\vec{f} = (x^3 - yz)i$ 2 $\vec{i} - 2x^2y\vec{j} + 2\vec{k}$ over the cube bounded by x = 0, y = 0, z = 0 and x = a, y = a, z = a.

Solution:

Gauss-divergence theorem states that:

$$\iint_S \vec{F}\cdot\hat{n}\,ds = \iiint_V (\nabla\cdot\vec{F})\,dv$$

$$\nabla\cdot\overline{F} = \frac{\partial f_1}{\partial x} + \frac{\partial f_2}{\partial y} + \frac{\partial f_3}{\partial z}$$

$$= 3x^2 - 2x^2 = x^2$$

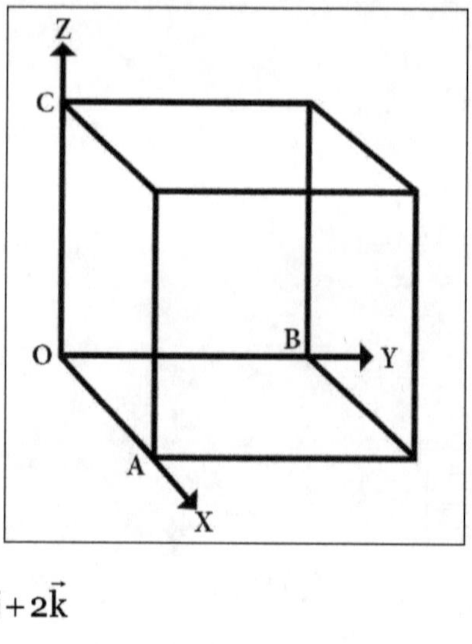

L.H.S. $\vec{F} = (x^3 - yz)i - 2x^2yj + 2\vec{k}$

$$\iint_S \overline{F}.\hat{n}\,ds = \iint_{S_1} + \iint_{S_2} + \iint_{S_3}$$

$$\begin{pmatrix} x = 0 \\ dx = 0 \\ ft = -i \\ dS = dy\,dz \end{pmatrix} \begin{pmatrix} x = a \\ dx = 0 \\ ft = -i \\ dS = dy\,dz \end{pmatrix} \begin{pmatrix} y = 0 \\ dy = 0 \\ ft = -j \\ dS = dx\,dz \end{pmatrix}$$

$$+ \iint_{s_4} \quad + \quad \iint_{s_5} \quad + \quad \iint_{s_6}$$

$$\begin{pmatrix} y = a \\ dy = 0 \\ ft = -j \\ dj = dx\,dz \end{pmatrix} \begin{pmatrix} z = 0 \\ dx = 0 \\ ft = -k \\ dS = dx\,dy \end{pmatrix} \begin{pmatrix} z = a \\ dz = 0 \\ ft = k \\ dS = dx\,dy \end{pmatrix}$$

$$= -\iint (x^3 - yz)\,dy\,dz + \iint (x^3 - yz)\,dy\,dz + \iint 2x^2 y\,dx\,dy$$

$$\begin{pmatrix} x = 0 \\ dx = 0 \end{pmatrix} \qquad \begin{pmatrix} x = a \\ dx = 0 \end{pmatrix} \qquad \begin{pmatrix} y = 0 \\ dy = 0 \end{pmatrix}$$

$$+\iint (-2x^2)\,dx\,dy \quad + \quad \iint (-2)\,dx\,dy \quad + \quad \iint (2)\,dx\,dy$$

$$\begin{pmatrix} y = a \\ dy = 0 \end{pmatrix} \qquad \begin{pmatrix} z = 0 \\ dz = 0 \end{pmatrix} \qquad \begin{pmatrix} z = a \\ dz = 0 \end{pmatrix}$$

$$= \int_0^a \int_0^a yz\,dy\,dz + \int_0^a \int_0^a (a^3 - yz)\,dy\,dz + 0$$

$$-2a \int_0^a \int_0^a x^2 dx\,dz - 2\int_0^a \int_0^a dx\,dy + 2\int_0^a \int_0^a dx\,dy$$

$$= \frac{a^4}{4} + a^5 - \frac{a^4}{4} - \frac{2a^5}{3} = \frac{a^5}{3} \qquad \ldots(I)$$

$$\text{R.H.S.} = \iiint_V \nabla \cdot \bar{F}\,dv = \int_0^a \int_0^a \int_0^a x^2 dx\,dy\,dz$$

$$= \int_0^a \int_0^a \left(\frac{x^3}{3}\right)_0^a dy'dz$$

$$= \left(\frac{a^3}{3}\right) \int_0^a (y)_0^a\,dz$$

$$= \left(\frac{a^4}{3}\right) \int_0^a dz = \left(\frac{a^4}{3}\right)(z)_0^a$$

$$= \frac{a^5}{3} \quad ...(II)$$

From (I) and (II)

$$\iint_s \bar{F}.\hat{n}\, ds = \iiint_v \nabla.\bar{F}\, dv = \frac{a^5}{3}$$

Hence Gauss-Divergence Theorem is verified.

2. Let us verify divergence theorem for $\bar{F} = x^2 i + z\bar{j} + yz\bar{k}$ over the cube formed by the planes x = ± 1, y = ± 1, z = ± 1.

Solution:

Gauss divergence theorem states that,

$$\iint_s \bar{F}\cdot\hat{n}\, ds. = \hat{n}\, ds = \iiint_v \nabla\cdot\bar{F}\, dv \quad ...(1)$$

Here, $\bar{F} = x^2\vec{i} + z\vec{j} + yz\vec{k}.$

$$\nabla\cdot\vec{F} = 2x + 0 + y = 2x + y$$

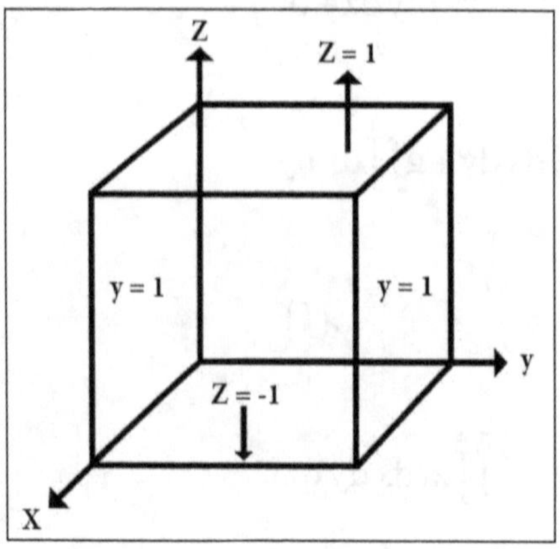

LHS:

$$\bar{F} = x^2\vec{i} + z\vec{j} + yz\vec{k}$$

$$\iint_s \bar{F}\cdot\hat{n}\, ds = \int_{S_1} \begin{pmatrix} x = -1 \\ dx = 0 \\ \hat{n} = -1 \\ ds = dy\, dz \end{pmatrix} + \int_{S_2} \begin{pmatrix} x = 1 \\ dx = 0 \\ \hat{n} = i \\ ds = dy\, dz \end{pmatrix} + \int_{S_3} \begin{pmatrix} y = -1 \\ dy = 0 \\ \hat{n} = -j \\ ds = dx\, dz \end{pmatrix}$$

$$\int_{S_4} \begin{pmatrix} y=1 \\ dy=0 \\ \hat{n}=j \\ ds=dx\,dz \end{pmatrix} + \int_{S_5} \begin{pmatrix} z=-1 \\ dz=0 \\ \hat{n}=-k \\ ds=dx\,dy \end{pmatrix} + \int_{S_6} \begin{pmatrix} z=-1 \\ dz=0 \\ \hat{n}=k \\ ds=dx\,dy \end{pmatrix}$$

$$= \int_{S_1} (-x^2)\,dy\,dz + \int_{S_2} (x^2)\,dy\,dz + \int_{S_3} (-z)\,dx\,dz$$

$$\quad x=-1 \qquad x=1 \qquad y=-1$$

$$+ \int_{S_4} z\,dx\,dz + \int_{S_5} (-yz)\,dx\,dy + \int_{S_6} yz\,dx\,dy$$

$$\quad y=1 \qquad z=-1 \qquad z=1$$

$$= -\int_{-1}^{1}\int_{-1}^{1} dy\,dz + \int_{-1}^{1}\int_{-1}^{1} dy\,dz - \int_{-1}^{1}\int_{-1}^{1} z\,dx\,dz$$

$$+ \int_{-1}^{1}\int_{-1}^{1} dx\,dz + \int_{-1}^{1}\int_{-1}^{1} y\,dx\,dy + \int_{-1}^{1}\int_{-1}^{1} y\,dx\,dy$$

$$= 2\int_{-1}^{1} (xy)_{-1}^{1}\,dy = 4\int_{-1}^{1} y\,dy$$

$$= 4\left(\frac{y^2}{2}\right)_{-1}^{1} = 0 \quad \ldots(\mathrm{I})$$

RHS:

$$\iiint_V \nabla\cdot\bar{F}\,dv = \int_{-1}^{1}\int_{-1}^{1}\int_{-1}^{1} (2x+y)\,dx\,dy\,dz$$

$$= \int_{-1}^{1}\int_{-1}^{1} \left[x^2 + xy\right]_{-1}^{1} dy\,dz$$

$$= \int_{-1}^{1}\int_{-1}^{1} \left[(1+y)-(1-y)\right] dy\,dz$$

$$= \int_{-1}^{1}\int_{-1}^{1} 2y\,dy\,dx$$

$$= \int_{-1}^{1} \left[y^2 \right]_{-1}^{1} = 0 \quad ...(II)$$

From (I) and (II), Divergence theorem is verified.

3. Let us verify Gauss's divergence theorem for $\vec{F} = 4xz\vec{i} - y^2 j + yz\vec{k}$ over the cube bounded by x = 0, x = 1, y = 0, y = 1, z = 0 and z = 1.

Solution:

Gauss divergence theorem states that:

$$\iint_S \overline{F} \cdot \hat{n}\, ds = \iiint_V \text{div}\ \overline{F} \cdot dv$$

Given: $\overline{F} = 4xzi - y^2 j + yz\vec{k}$

$$\nabla \cdot \overline{F} = \frac{\partial f_1}{\partial x} + \frac{\partial f_2}{\partial x} + \frac{\partial f_3}{\partial z} = 4z - 2y + y = 4z - y$$

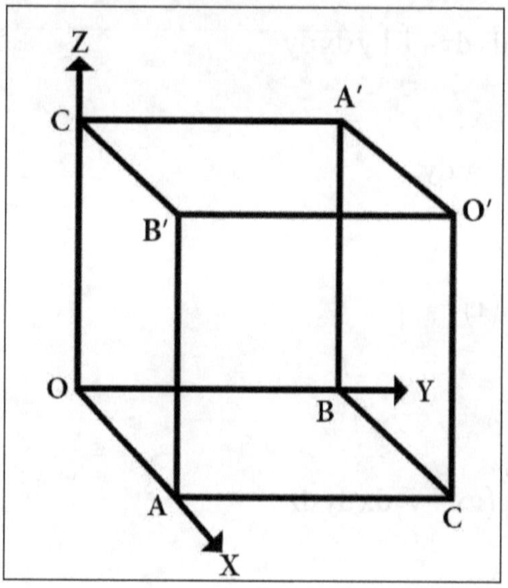

L.H.S: $F = 4xzi - ys\, j + yzk^{-1}$

$$\iint_S \overline{F} \cdot \hat{n}\, ds = \iint_{S_1} \overline{F} \cdot \hat{n}\, ds + \iint_{S_2} + \iint_{S_3}$$

$$\begin{pmatrix} x = 0 \\ dx = 0 \\ \hat{n} = -i \\ ds = dy\,dz \end{pmatrix} \quad \begin{pmatrix} x = 0 \\ dx = 0 \\ \hat{n} = i \\ ds = dy\,dz \end{pmatrix} \quad \begin{pmatrix} y = 0 \\ dy = 0 \\ n = -j \\ ds = dy\,dz \end{pmatrix}$$

$$+\iint\limits_{S_4} + \iint\limits_{S_5} + \iint\limits_{S_6}$$

$$\begin{pmatrix} y=1 \\ dy=0 \\ \hat{n}=j \\ ds=dxdz \end{pmatrix} \begin{pmatrix} z=0 \\ dz=0 \\ \hat{n}=-k \\ ds=dxdy \end{pmatrix} \begin{pmatrix} z=1 \\ dz=0 \\ \hat{n}=k \\ ds=dxdy \end{pmatrix}$$

$$=\iint(-4xz)\,dy\,dz+\iint 4xz\,dy\,dz+\iint y^2\,dx\,dz$$

$$\begin{pmatrix} x=0 \\ dx=0 \end{pmatrix} \begin{pmatrix} x=1 \\ dx=0 \end{pmatrix} \begin{pmatrix} y=0 \\ dy=0 \end{pmatrix}$$

$$+\iint\left(-y^2\right)dx\,dz+\iint(-yz)\,dx\,dy+\iint dx\,dy$$

$$\begin{pmatrix} y=1 \\ dy=0 \end{pmatrix} \begin{pmatrix} z=0 \\ dz=0 \end{pmatrix} \begin{pmatrix} z=1 \\ dz=0 \end{pmatrix}$$

$$\iint\limits_S \bar{F}\cdot\hat{n}\,ds=0+\int_0^1\!\!\int_0^1 4z\,dy\,dz+0-\int_0^1\!\!\int_0^1 dx\,dz+0+\int_0^1\!\!\int_0^1 y\,dx\,dy$$

$$=2-1+\frac{1}{2}=\frac{3}{2} \quad \ldots(I)$$

R.H.S:

$$\iiint\limits_V \nabla\cdot\bar{F}dv=\int_0^1\!\!\int_0^1\!\!\int_0^1(4z-y)\,dx\,dy\,dz$$

$$=\int_0^1\!\!\int_0^1\left[4xz-xy\right]_0^1 dy\,dz$$

$$=\int_0^1\!\!\int_0^1\left[4z-y\right]dy\,dz$$

$$=\int_0^1\left[4yz-\frac{y^2}{2}\right]_0^1 dz$$

$$=\int_0^1\left[4z-\frac{1}{2}\right]_0^1 dz$$

$$= \left[2z^2 - \frac{z}{2}\right]_0^1 = 2 - \frac{1}{2}$$

$$= \frac{3}{2} \quad ...(II)$$

From (I) & (II)

$$\iint_S \bar{F}\cdot\hat{n}\,ds = \iiint_V \nabla\cdot\bar{F}\,dv = \frac{3}{2}$$

Hence Gauss Divergence Theorem is verified.

Stoke's Theorem

Statement: If S is an open surface bounded by a simple closed curve C and if a vector Function $\vec{F}$ is continuous and has continuous partial derivatives in S and on $\iint \text{curlI}\,\vec{F}\cdot\vec{n}\,ds = \int \vec{F}\cdot d\vec{r}\ \hat{n}$ where $\hat{n}$ is the unit vector normal to the surface (i.e) The surface integral of the normal component of curl $\vec{F}$ is equal to the integral of the tangential component of $\vec{F}$ taken around C.

Problems

1. By using Stokes theorem let us prove that curl grad φ = 0.

Solution:

Stokes theorem gives:

$$\int_C \bar{f}.\bar{d}r = \iint_S \left(\text{curl}\,\bar{f}\right)\cdot\vec{n}\,dS.$$

Let $\quad \bar{f} = \nabla\phi = \hat{i}\frac{\partial\phi}{\partial x} + \hat{j}\frac{\partial\phi}{\partial y} + \hat{k}\frac{\partial\phi}{\partial z}$

$$(\nabla\phi)\cdot d\bar{r} = \frac{\partial\phi}{\partial x}dx + \frac{\partial\phi}{\partial y}dy + \frac{\partial\phi}{\partial z}dz = d\phi$$

$$\therefore \iint_S \left(\text{curl}\,\bar{f}\right)\cdot\hat{n}\,ds = \iint_S \left[\text{curl}(\nabla\phi)\right]\cdot\hat{n}\,ds$$

$$= \int_C (\nabla\phi)\cdot d\bar{r}$$

$$= \int_C d\phi$$

$= c$

$\therefore$ (curl f) grad $\varphi = 0$.

2. By using Stokes theorem, let us evaluate $\int_C \vec{F}.\vec{d}r$, where $\vec{F} = y^2\vec{i} + x^2\vec{j} - (x+z)\vec{k}$ and 'C' is the boundary of the triangle with vertices at (0, 0, 0), (1, 0, 0), (1, 1, 0).

Solution:

Stokes Theorem gives:

$$\int_C \overline{F} \cdot d\overline{r} = \iint_S (\nabla \times \overline{F}) \cdot \hat{n}\, ds \qquad \text{... 1} \qquad \text{...(1)}$$

$$\overline{F} = y^2\,\hat{i} + x^2\,\hat{j} - (x+z)\hat{k}$$

$$\nabla \times \overline{F} = \begin{vmatrix} \hat{i} & \hat{j} & \hat{k} \\ \partial/\partial x & \partial/\partial y & \partial/\partial z \\ y^2 & x^2 & -(x+z) \end{vmatrix}$$

$$= \hat{i}(0-0) - \hat{j}(-1-0) + \hat{k}(2x-2y)$$

$$= \hat{j} + 2(x-y)\hat{k}$$

Let $\quad I = \int_C \overline{F} \cdot d\overline{r}$

$$= \iint_S (\nabla \times \overline{F}) \cdot \hat{n}\, ds$$

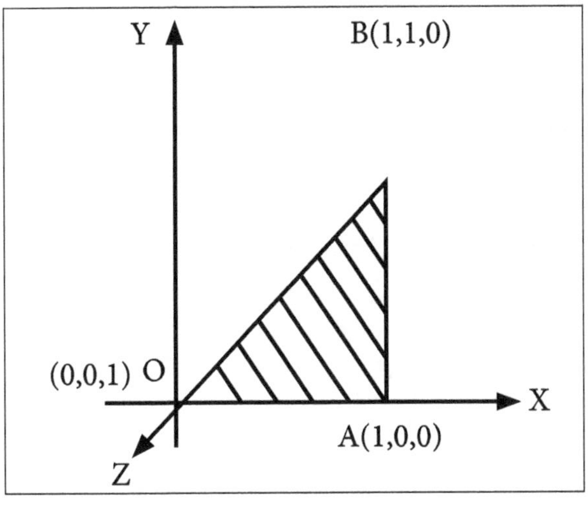

S is triangle in XY Plane:

$$\hat{n} = \hat{k}$$

$$(\nabla \times \overline{F}) \cdot \hat{n} = 2(x - y)$$

$$ds = \frac{dxdy}{|\hat{n} \cdot \hat{k}|} = dx\,dy$$

Limits are:

x : 0 → 1Equation of OB is y = x.

y : 0 → x Equation of AB is x = 1.

Equation of AB is x = 1:

$$\therefore I = \int_0^1 \int_0^x 2(x - y)\,dy\,dx$$

$$= 2\int_0^1 \left[xy - \frac{y^2}{2} \right]_0^x dx = 2\int_0^1 \left[x^2 - \frac{x^2}{2} \right] dx$$

$$= 2\int_0^1 \frac{x^2}{2} dx = \left[\frac{x^3}{3} \right]_0^1 = \frac{1}{3}$$

3. Let us verify Stokes theorem for the vector field $\overline{F} = (2x - y)\overline{i} - yz^2\overline{j} - y^2z\overline{k}$ Over the upper half surface $x^2 + y^2 + z^2 = 1$, bounded by its projection on the xy-plane.

Solution:

Stokes theorem states that:

$$\int_C \overline{F} \cdot \overline{dr} = \iint_S \text{curl}\,\overline{F} \cdot \hat{n}\,ds$$

$$\overline{F} \cdot \overline{dr} = (2x - y)dx - yz^2dy - y^2zdz$$

$$\text{curl } \overline{F} = \begin{vmatrix} i & j & k \\ \frac{\partial}{\partial x} & \frac{\partial}{\partial y} & \frac{\partial}{\partial z} \\ 2x - y & -yz^2 & -y^2 \end{vmatrix}$$

$$= \bar{i}\left(-2yz+2yz\right)-j(0)+k(0+1) = \bar{k}$$

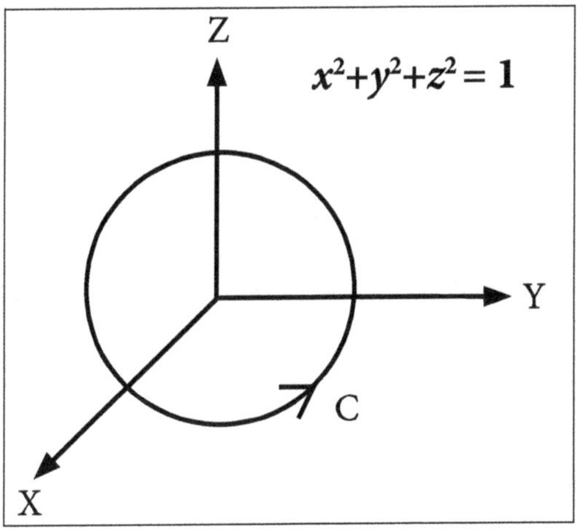

LHS:

$$\int \bar{F}\cdot \bar{d}r = \int (2x-y)dx$$

Put $x = \cos\theta$; $y = \sin\theta \Rightarrow dx = -\sin\theta\, d\theta$

θ varies from 0 to 2π:

$$\int_C \bar{F}.\ dr = \int_0^{2\pi} [2\cos\theta - \sin\theta](-\sin\theta d\theta)$$

$$= \int_0^{2\pi} \left[-2\sin\theta\cos\theta + \sin^2\theta\right]d\theta$$

$$= \int_0^{2\pi} \left[-\frac{\sin 2\theta}{1} + \frac{1}{2} - \frac{\cos 2\theta}{2}\right]d\theta$$

$$= \left[\frac{\cos 2\theta}{2} + \frac{\theta}{2} - \frac{\sin 2\theta}{4}\right]_0^{2\pi}$$

$$= \left[\left(\frac{1}{2}+\pi\right) - \left(\frac{1}{2}\right)\right] = \pi \quad ...(I)$$

RHS:

$$\iint_S \text{curl}\,\bar{F}\cdot\hat{n}\,ds = \iint \vec{k}\cdot\left(\bar{k}\right)dx\,dy$$

$$= \iint_R dx\,dy$$

$$= \text{Area of the Circle } x^2 + y^2 = 1$$

$$= \pi \qquad ...(II)$$

From (I) and (II), Stokes theorem is verified.

4. Let us verify Stokes theorem for $\bar{F} = (x^2 + y^2)\bar{i} - 2xy\bar{j}$ taken around the rectangle formed by the lines x = - a, x = + a, y = 0 and y = b.

Solution:

Stokes theorem states that:

$$\int_C \bar{F}\cdot\bar{dr} = \iint_S \text{curl } \bar{F}\cdot n\,ds$$

$$\bar{F}.\bar{dr} = (x^2 + y^2)dx - 2xydy$$

C is enclosed by x = - a ; x = a , y = 0 & y = b.

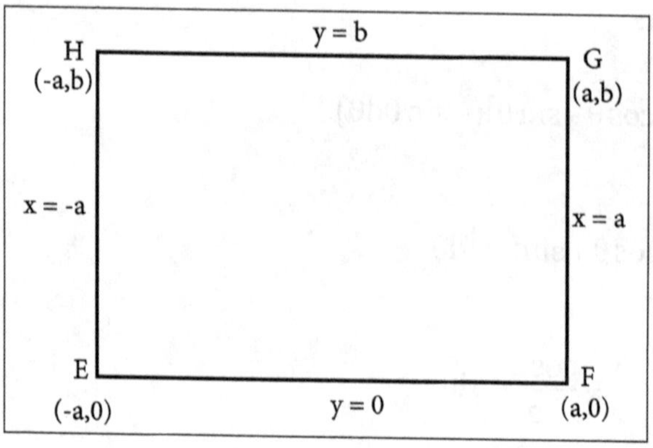

LHS:

$$\int_C \bar{F}\cdot\bar{dr} = \int_C \left[(x^2 + y^2)dx - 2xy\,dy\right]$$

$$= \int_{\begin{pmatrix} EF \\ y=0 \\ dy=0 \end{pmatrix}} + \int_{\begin{pmatrix} FG \\ x=a \\ dx=0 \end{pmatrix}} + \int_{\begin{pmatrix} GH \\ y=b \\ dy=0 \end{pmatrix}} + \int_{\begin{pmatrix} HE \\ x=-a \\ dx=0 \end{pmatrix}}$$

$$= \int_{-a}^{a} x^2 dx + \int_{0}^{b} (-2ay dy) + \int_{a}^{-a} (x^2 + b^2) dx + \int_{b}^{0} 2ay\, dy$$

$$= \left(\frac{x^3}{3}\right)_{-a}^{a} - \left(ay^2\right)_0^b + \left(\frac{x^3}{3} + b^2 x\right)_a^{-a} + \left(ay^2\right)_b^0$$

$$= \left(\frac{a^3}{3} + \frac{a^3}{3}\right) - ab^2 + \left(-\frac{a^3}{3} - ab^2 - \frac{a^3}{3} - ab^2\right) - ab^2$$

$$\text{LHS} = -4ab^2 \qquad \text{...(1)}$$

RHS of Stokes theorem:

$$\text{curl}\,\overline{F} = \nabla \times \overline{F} = \begin{vmatrix} \vec{i} & \vec{j} & \vec{k} \\ \frac{\partial}{\partial x} & \frac{\partial}{\partial y} & \frac{\partial}{\partial z} \\ x^2 + y^2 & -2xy & 0 \end{vmatrix}$$

$$= \vec{i}(0) - \vec{j}(0) + \vec{k}(-2y - 2y)$$

$$= -4y\vec{k}$$

$$\iint_S \text{curl}\,\overline{F} \cdot \hat{n}\, ds = \iint_R \left(-4y\vec{k}\right) \cdot \vec{k}\, dx\, dy$$

$$= -4\int_0^b \int_{-a}^{a} y\, dx\, dy$$

$$= -4\int_0^b (yx)_{-a}^{a}\, dy$$

$$= -4\int_0^b [ay + ay]dy$$

$$= -8a\int_0^b y\, dy = -8a\left(\frac{y^2}{2}\right)_0^b$$

$$= -4ab^2 \qquad \text{...(2)}$$

From (1) and (4):

$$\int_C \overline{F}\cdot\overline{d}r = \iint_S \text{curl } \overline{F}\cdot\hat{n}ds = -4ab^2.$$

Hence Stokes theorem is verified.

5.3 Solenoidal Vectors, Irrotational Vectors and Finding Potential Function

Irrotational Vector

A vector $\vec{F}$ is called irrotational if Curl $\vec{F} = 0$ (i.e) if $\nabla\times\vec{F}=0$.

Scalar Potential

If $\vec{F}$ is an irrotational vector, then there exists a scalar function φ Such that $\vec{F}=\nabla\phi$. Such a scalar function is called scalar potential of $\vec{F}$.

Solenoidal Vector

A vector $\vec{F}$ is said to be solenoidal if div $\vec{F}$ =0 (i.e) $\nabla\times\vec{F}=0$.

Problems

1. Let us prove that $\vec{F}=(6xy+z^3)\vec{i}+(3x^2-z)\vec{j}+(3xz^2-y)\vec{k}$ is irrotational vector and find the scalar potential such that $\vec{F}=\nabla\phi$

Solution:

If $\nabla\times\overline{F}=0$ then $\vec{F}$ is irrotational.

Now:

$$\nabla\times\overline{F}=\begin{vmatrix} i & j & k \\ \frac{\partial}{\partial x} & \frac{\partial}{\partial y} & \frac{\partial}{\partial z} \\ 6xy+z^3 & 3x^2-2 & 3xz^2-y \end{vmatrix}$$

$$=i(-1+1)-j(3z^2-3z^2)+k(6x-6x)$$

$$=0$$

Here $\overline{F}$ is irrotational.

$$\nabla\phi = \overline{F}$$

$$\text{i.e., } \overline{i}\frac{\partial\phi}{\partial x} + \overline{j}\frac{\partial\phi}{\partial y} + \overline{k}\frac{\partial\phi}{\partial z} = \left(6xy + z^3\right)\overline{i} + \left(3x^2 - z\right)\overline{j} + \left(3xz^2 - y\right)\overline{k}$$

Equating like terms:

$$\text{i.e., } \frac{\partial\phi}{\partial x} = 6xy + z^3 \quad ...(i)$$

Integrating:

$$\phi = 3x^2y + xz^3 + f_1(y,z) \qquad ...(1)$$

$$\frac{\partial\phi}{\partial y} = 3x^2 - z \quad ...(ii)$$

Integrating:

$$\phi = 3x^2y - yz + f_2(x,z) \qquad ...(2)$$

$$\frac{\partial\phi}{\partial z} = 3xz^2 - y \quad ...(iii)$$

Integrating:

$$\phi = xz^3 - yz + f_3(x,y) \qquad ...(3)$$

From (1), (2) and (3):

$$\phi = (x,y,z) = 3x^2y + xz^3 - yz + C$$

2. Let us show that $\vec{F} = \left(y^2 + 2xz^2\right)\vec{i} + \left(2xy - z\right)\vec{j} + \left(2x^2z - y + 2z\right)\vec{k}$ is irrotational and hence and its scalar potential.

Solution:

To prove $\overline{F}$ is irrotational.

i.e., to prove $\nabla \times \overline{F} = 0$.

$$\text{LHS} = \nabla \times \overline{F} = \begin{vmatrix} \hat{i} & \hat{j} & \hat{k} \\ \partial/\partial x & \partial/\partial y & \partial/\partial z \\ y^2 + 2xz^2 & 2xy - z & 2x^2z - y + 2z \end{vmatrix}$$

$$=\hat{i}\left[\partial/\partial y\left(2x^2z-y+2z\right)-\partial/\partial z\left(2xy-z\right)\right]-\hat{j}\left[\frac{\partial}{\partial x}\left(2x^2z-y+2z\right)-\frac{\partial}{\partial z}\left(y^2+2xz^2\right)\right]$$

$$+\hat{k}\left[\frac{\partial}{\partial x}(2xy-z)-\frac{\partial}{\partial y}\left(y^2+2xz^2\right)\right]$$

$$=\hat{i}\left[-1-(-1)\right]-\hat{j}[4xz-4xz]+\hat{k}[2y-2y]$$

$$=0\hat{i}+0\hat{j}+0\hat{k}$$

$$=0\ \forall x,y,z$$

To find scalar potential of $\bar{\ }$:

Let $\bar{F}=\nabla_\phi$ then φ is scalar potential of $\bar{F}$.

$$\therefore \bar{F}=\hat{i}\frac{\partial\phi}{\partial x}+\hat{j}\frac{\partial\phi}{\partial y}+\hat{k}\frac{\partial\phi}{\partial z}$$

$$\Rightarrow\frac{\partial\phi}{\partial x}=y^2+2xz^2\Rightarrow\phi=xy^2+x^2z^2$$

$$\frac{\partial\phi}{\partial y}=2xy-z\Rightarrow\phi=xy^2-yz$$

$$\frac{\partial\phi}{\partial z}=2x^2z-y+2z\Rightarrow\phi=x^2z^2-yz+z^2$$

$$\phi=xy^2+x^2z^2-yz+z^2+c$$

Is required scalar potential of $\bar{F}$.

3. Let us show that $\bar{F}=\left(2xy-z^2\right)\bar{i}+\left(x^2+2yz\right)\bar{j}+\left(y^2-2zx\right)\bar{k}$ is irrotational and find its scalar potential.

Solution:

If $\nabla\times\bar{F}=0$ then $\bar{F}$ is irrotational.

$$\nabla\times\bar{F}=\begin{vmatrix}\bar{i} & \bar{j} & \bar{k}\\ \frac{\partial}{\partial x} & \frac{\partial}{\partial y} & \frac{\partial}{\partial z}\\ 2xy-z^2 & x^2+2yz & y^3-2zx\end{vmatrix}$$

$$=\vec{i}[2y-2y]-\vec{j}(-2z)-(-2z)+\vec{k}(2x-2x)$$

$$=\vec{o}$$

Hence $\overline{F}$ is irrotational.

If $\overline{F}$ is irrotational then $\nabla\phi=\overline{F}$

i.e., $\vec{i}\frac{\partial\phi}{\partial x}+\vec{j}\frac{\partial\phi}{\partial y}+k\frac{\partial\phi}{\partial z}$

$$=\left(2xy-z^2\right)\vec{i}+\left(x^2+2yz\right)\vec{j}\left(y^2-2zx\right)\vec{k}$$

$$\therefore\frac{\partial\phi}{\partial x}=2xy-z^2 \quad ...(1)$$

$$\frac{\partial\phi}{\partial y}=x^2+2yz \quad ...(2)$$

$$\frac{\partial\phi}{\partial z}=y^2-2zx \quad ...(3)$$

Integrating (1), (2) and (3):

$$\phi=x^2y-xz^2+f(y,z) \quad ...(4)$$

$$\phi=x^2y+y^2z+f(x,z) \quad ...(5)$$

$$\phi=y^2z-xz^2+f(x,y) \quad ...(6)$$

From (4), (5) and (6) we get:

$$\phi(x,y,z)=x^2y-xz^2+y^2z+C$$

4. Let us show that the vector field $\overline{F}=\left(x^2+xy^2\right)\overline{i}+\left(y^2+x^2y\right)\overline{j}$ is irrotational and find its scalar potential.

Solution:

$\nabla\times\overline{F}=0$ then $\overline{F}$ is said to be irrotational.

$$\nabla\times\overline{F}=\begin{vmatrix}\vec{i} & \vec{j} & \vec{k}\\ \frac{\partial}{\partial x} & \frac{\partial}{\partial y} & \frac{\partial}{\partial z}\\ x^2+xy^2 & y^2+x^2y & 0\end{vmatrix}$$

$$=\vec{i}[o]-\bar{j}[o]+\bar{k}[2xy-2xy]=\bar{o}$$

$\therefore \bar{F}$ is irrotational

If $\bar{F}$ is irrotational then $\nabla\phi \therefore F$

$$\vec{i}\frac{\partial\phi}{\partial x}+j\frac{\partial\phi}{\partial y}+k\frac{\partial\phi}{\partial z}=\left(x^2+xy^2\right)\vec{i}+\left(y^2+x^2y\right)\bar{j}$$

$$\Rightarrow\frac{\partial\phi}{\partial x}=x^2+xy^2;\ \frac{\partial\phi}{\partial y}=y^2+x^2y$$

$$\Rightarrow\phi=\frac{x^3}{3}+\frac{x^2y^2}{2}+f(y,z);\phi=\frac{y^3}{3}+\frac{x^2y^2}{2}+f(x,z)$$

$$\therefore\phi=\frac{x^3}{3}+\frac{x^2y^2}{2}+\frac{y^3}{3}+C$$

Vector identities - div(ϕA), curl (ϕA), curl(grad ϕ), div(curl A).

Permissions

All chapters in this book are published with permission under the Creative Commons Attribution Share Alike License or equivalent. Every chapter published in this book has been scrutinized by our experts. Their significance has been extensively debated. The topics covered herein carry significant information for a comprehensive understanding. They may even be implemented as practical applications or may be referred to as a beginning point for further studies.

We would like to thank the editorial team for lending their expertise to make the book truly unique. They have played a crucial role in the development of this book. Without their invaluable contributions this book wouldn't have been possible. They have made vital efforts to compile up to date information on the varied aspects of this subject to make this book a valuable addition to the collection of many professionals and students.

This book was conceptualized with the vision of imparting up-to-date and integrated information in this field. To ensure the same, a matchless editorial board was set up. Every individual on the board went through rigorous rounds of assessment to prove their worth. After which they invested a large part of their time researching and compiling the most relevant data for our readers.

The editorial board has been involved in producing this book since its inception. They have spent rigorous hours researching and exploring the diverse topics which have resulted in the successful publishing of this book. They have passed on their knowledge of decades through this book. To expedite this challenging task, the publisher supported the team at every step. A small team of assistant editors was also appointed to further simplify the editing procedure and attain best results for the readers.

Apart from the editorial board, the designing team has also invested a significant amount of their time in understanding the subject and creating the most relevant covers. They scrutinized every image to scout for the most suitable representation of the subject and create an appropriate cover for the book.

The publishing team has been an ardent support to the editorial, designing and production team. Their endless efforts to recruit the best for this project, has resulted in the accomplishment of this book. They are a veteran in the field of academics and their pool of knowledge is as vast as their experience in printing. Their expertise and guidance has proved useful at every step. Their uncompromising quality standards have made this book an exceptional effort. Their encouragement from time to time has been an inspiration for everyone.

The publisher and the editorial board hope that this book will prove to be a valuable piece of knowledge for students, practitioners and scholars across the globe.

Index

www.ingramcontent.com/pod-product-compliance
Lightning Source LLC
LaVergne TN
LVHW080637210926
848485LV00003B/281